FACES OF MEDICINE

FACES
OF MEDICINE

A PHILOSOPHICAL STUDY

by

Wim J. van der Steen

Free University, Amsterdam,
and University of Leiden, The Netherlands

and

P. J. Thung

University of Leiden,
and University of Limburg, Maastricht, The Netherlands

Springer-Science+Business Media, B.V.

Library of Congress Cataloging in Publication Data

Steen, Wim J. van der, 1940-
    Faces of medicine : a philosophical study / Wim J. van der Steen
and P.J. Thung.
        p.   cm.
    Bibliography: p.
    Includes index.
    ISBN 978-94-010-7124-6   ISBN 978-94-009-1397-4 (eBook)
    DOI 10.1007/978-94-009-1397-4

    1. Medicine--Philosophy.   I. Thung, P. J., 1924-   . II. Title.
R723.S76 1988
362.1'01--dc19   87-37374
                                                              CIP

CONTENTS

CHAPTER I. INTRODUCTION

1. MEDICINE

Illness, disease and disability plague man in every culture. But the form they take is not the same everywhere. Neither is man's reaction. Coping strategies, and the experience and knowledge backing them, depend very much on cultural setting. So medicine, the fabric of strategy and knowledge, can only be understood in the context of culture.

In western society today, severe judgements are passed on medicine. Its store of knowledge and experience, and its repertory of strategies, have grown immensely during the last few decades. But it hardly alleviates dominant ailments, especially chronic diseases, diseases of old age and disturbances of social and mental functioning.

We know that these ailments have come to the fore as the incidence of more "primitive" diseases declined in industrial societies. Infant deaths, and malnutrition and infections striking at young age, have dwindled to marginal significance in Western Europe and life expectancy at birth is twice that of some 150 years ago. Thus our new troubles are connected with past successes.

Modern civilization came with hygienic and prosperous ways of life which in the 19th century disposed of large-scale epidemics. It gave us vaccinations and other preventive measures as well as insight into dietary requirements, which prolonged average life-span far beyond the age of biological fecundity. There are profound consequences, biologically and culturally. Increasingly, we fall prey to ills of old age. And, in the aftermath of prosperity, we live densely packed together in an environment which exudes the psychological and physical hazards we created. What about health, disease and illness in this setting? It has been thought that they should be approached with more scientific knowledge and techniques, but we clearly need much more than that.

How do we meet the problems of health and disease which confront us today? So far, medicine in our society has concentrated on science

1

(especially biology) and technology to master the dominant "western" diseases. Fundamental research, sophisticated diagnostic tools and therapies, and an extensive pharmacological repertory, all exploit the somatological approach to these diseases. Epidemiology and psychology do show that behaviour has its share in pathogenesis, but knowledge of behaviour is hardly translated into practical measures. Laymen and professionals alike are preoccupied by somatological conceptions. Institutional and industrial health care interests are biased in the same way. All in all, with traditional medicine modernized, we are apparently engaged in a costly but loosing battle against unprecedented odds. Medicine is equipped to combat many well-defined environmental hazards threatening specific groups or individuals. It may also repair or alleviate damage incurred by individuals, especially if it is acute. But it is not equipped to prevent or cure all the chronic ailments, many of them interrelated and vaguely defined, that are associated with modern life. That is where criticism of medicine sets in. The purpose of this book is to uncover philosophical issues behind this criticism as well as philosophical presuppositions of medicine itself.

Medicine is said to be both a science and a practice. What would the locus be of present inadequacies of medicine? Is our knowledge inadequate and is that why we fail in practice? Or don't we live up to scientific insights, so that we fail in the practical application of our knowledge? Or, finally, is there a mismatch between science and practice so that the two are not pursuing common goals? Such questions are fundamental to philosophical understanding of medicine.

Present failures of medicine are often attributed to incomplete knowledge. Physiological mechanisms at the basis of many modern diseases are as yet poorly known. The common suggestion is then that more research, along conventional scientific lines, will eventually yield remedies. The tools are at hand and our ultimate victory over diseases like cancer, rheumatoid arthritis and multiple sclerosis is just a matter of patient perseverance. But is this a realistic view? And if it is, would not additional successes rather increase suffering as ailments increase with age?

Research efforts in regular medicine which optimistically aim at "solving the riddle" of difficult diseases, are inspired by one and the same conceptual model. A disease is a set of functional and structural changes in the body which are elicited by specific causal agents or events, the identification of which is the key to prevention or therapy. In the approach to, say, infectious diseases, the model works reasonably well in theory and in practice. But there are signs that the diseases of modern civilization are different. No simple model really comes to grips with all forms of cancer. Even complex models may not work as long as one views human life through a biological window. If one continually tries to solve problems in a single way, one will be left with the problems that cannot be solved in that way! That is a truism, an important one we think.

Lack of success, in other cases, is blamed on practice rather than on theory. Tuberculosis, in theory, is not a problem anymore. Its microbial cause is known and the disease is easily detected and combated with immunization and chemotherapy. But in practice the disease still is a major problem in many countries. That is mostly regarded as a matter of practical short-comings. Where medical facilities and economic conditions are insufficient to provide for the detection, isolation and treatment of active cases, the disease will continue to spread, especially among the poor. Medical science has a solution, but it is not implemented.

However, one can also look at the situation in a different way. From a social point of view, tuberculosis belongs to the conditions of life in areas of poverty and underdevelopment. Its ravages there are part of the pattern of undernourishment, inadequate housing, economic vulnerability and medical neglect. From this perspective, conventional medical science is almost irrelevant.

The example of tuberculosis is analogous to other cases, more relevant to western society. Medical science will oppose prevailing smoking habits which increase the incidence of bronchial carcinoma. But mere opposition informed by medical knowledge will not suffice to effectuate preventive measures in practice. Smoking is embedded in the social reality of large sections of our population. It belongs to rituals of disco-bars and many other social events, and it is stimulated by movies and advertisements.

Medicine often aims at its targets with great precision. The question is, does it aim at the right targets?

Theory and practice may be at odds in yet another way. Medical science, though optimistic about the future, will grant that the treatment of many types of disseminated cancer is now problematic. Chemotherapy may result in a temporary remission, but for how long, and how often, and at what costs of somatic and mental suffering? However, medicine even manages to cover such uncertainties with the rigour of science. The behaviour of disseminated cancer can be analysed in terms of quantitative probabilities. If my cancer is of such and such a type and stage, science may infer my expected life expectancy under non-treatment and under various kinds of treatment. On the basis of such analyses, a rational choice is offered. What do I prefer, given quantified alternatives and allied prices in terms of suffering and invalidity?

Now suppose that, during months of illness and fading hope, I have grown used to the idea that cancer is my final disease and that death is near. Suppose my views of life and death, shaped during my personal history, are such that I am now ready to accept this ending. Then the very suggestion of futher postponement through chemotherapy may threaten my painfully gained mental rest. The "rational choice" of medical science may often be irrelevant, even destructive, in the context of personal experience as a dimension of medical practice.

The problems described so far all concern fundamental relations between medicine and culture. Specifically, they call for a demarcation of medicine with respect to the problems of our time. Perhaps there are many problems with health which it cannot solve. Then one will need to be clear about that. Or perhaps it can solve more problems than it does. Then it may have to change.

In chapter II we will analyse these issues by evaluating three different views of medicine. We intend to show that extant general views very much depend on cultural setting. This might make one a relativist, but pervasive relativism is not our option. Once cultural bias is uncovered, there is a way to criticize it. We will advocate a pragmatic attitude which may help to pave the way for a more sensible medicine.

Shortcomings of modern medicine are visible. But it is not easy to uncover their source. It has been argued that the underlying scientific world-view is at fault. Science and allied technology are great, but they may not always be the proper tools for solving problems of illness and disease. Sometimes one may need alternative medicine with different sources of inspiration.

Indeed, in western societies there is a surge of interest in alternative ways of healing and in the concomitant alternative views of health and disease. This makes an evaluation of current trends in medicine even more difficult. Conventional medicine and some types of alternative medicine represent competing views of reality. So one cannot simply make one form of medicine the criterion to judge an alternative. If question begging is to be avoided one will have to deal with underlying philosophies. That is what we will do in chapter III.

The problem of medicine's inadequacy is but one symptom of a general uneasiness in modern culture. "Modernity", in the last two centuries, has overthrown so many basic notions about man, human society and human destiny, that practically all our ancient cultural coping mechanisms have withered away. The religious and moral foundations of society have been eroded and the ensuing struggle has not led to a new cultural coherence. In many places, there is a revival of religious fundamentalism. Elsewhere esoteric creeds are embraced as a reaction to moral uncertainties. The basic questions are of course perennial: what is man, what is his destination? But traditional approaches to these questions tend to ring false in modern ears and, more importantly, they do not deliver practical prescriptions for modern predicaments.

In medicine, too, there is a basic need for a "definition" of man. The ideal of medicine is said to be the promotion of health. That seems to pre-suppose a view of man, but there is no common view. Health is an elusive notion.

How can one define physical well-being, once one realizes that basic needs depend to a marked extent on cultural setting?

How can one come to grips with mental well-being as medical science and practice seem to account for human being mainly in terms of physical processes?

How can one agree on social well-being as different societies, even within the industrialized part of the world, are clearly built on different value-systems?

The first question addresses old problems with the definition of health and disease. Philosophers of medicine of all times have tried their hand at this question. They have mostly taken their cues from biology. However, many philosophers nowadays argue that the answer must be sought elsewhere. The very notions of health and disease presuppose values, and biology allegedly cannot deal with values. The issue remains contentious. It involves fundamental questions concerning the nature and goals of medicine. We will discuss it in chapter IV.

The other questions we formulated more overtly concern the "definition" of man. They are associated with a perennial philosophical problem which is mostly called the mind-body problem. We will analyse it in a general way in chapter V. Lastly, in chapter VI, we will consider the role of the mind-body problem in medicine. In doing that, we will emphasize limitations of *scientific* medicine. More specifically, the current emphasis on biology cannot be justified anymore.

2. PHILOSOPHY

If one wants to understand medicine, one will have to ask general questions *about* it which do not belong to medicine itself. That is where philosophy and other disciplines like history and sociology come in. Ours will mostly be a philosophical approach, a variety of approaches in fact. Philosophical affairs can be conducted in different ways, and there is no single one which suits sundry purposes.

In philosophy there is no agreement even on the nature of philosophy. How could one characterize it? Firstly, one could argue that philosophy is concerned with general questions which are left over by science either because it cannot solve them or because they do not belong to its domain (cf. "metaphysics", "ontology"). That is a negative characterization, but it would not be easy to formulate a positive one. The problem is that science also aims at generality, and there is no *a priori* limit to the questions it can answer. History has shown that boundaries between philosophy and science shift continually.

Secondly, there is the realm of normative matters (values and norms) which apparently belong to philosophy rather than science (cf. "ethics"). Would that provide a basis for a criterion of demarcation for philosophy *versus* science? To some extent it would. *In our culture* science is mostly concerned with empirical matters, it is supposed to be value-free. Unlike philosophy, it does not aim at laying foundations for normative views. However, it surely has a role in normative settings. When one aims at diminishing undernourishment in poor countries, one will need science in discussions about *means*. Moreover, the business of science includes the *empirical* study of values and norms. All in all, it would not be easy to formulate a *very* sharp criterion of demarcation in terms of normative issues.

Thirdly, as we said before, philosophy seems to be a place where one finds views *about* science (cf. "philosophy of science", specifically "methodology of science"). Would that help us to define philosophy? Again there is no easy answer. History and sociology also have theories about science, and there is disagreement about a proper division of labour. And there is yet another problem of demarcation. Science itself needs to reflect on science! Consider. Theories of medical science about disease can be adequate only if the *concept* of disease is sufficiently clear, and if statements about disease meet criteria like testability. So scientists will have to work at two different levels. It is their business to look at things and processes *and*, at a "higher level", at theories about things and processes. Methodology therefore does not exclusively belong to philosophy.

We think that the three items mentioned do characterize areas of philosophy, however vaguely. Now different schools of philosophy often emphasize different items. That may almost make them incomparable. None the less we will try, throughout this book, to combine divergent philosophical approaches to medicine. Anglo-Saxon philosophy of science and European phenomenology will get most attention.

Would it be possible to formulate a general strategy for dealing with health and disease? We have already argued in the previous section that the approach of modern medicine is problematic. Could philosophy help us to develop a better perspective? Part of the problem is doubtless that the *concepts* of health and disease are not very clear. So an analysis of the concepts in the style of Anglo-Saxon philosophy of science could be helpful. Philosophers of medicine in the U.S. have indeed produced a great amount of literature in which the concepts are analysed. But they have not reached consensus. There are those who would like to define "health" and "disease" in terms of plain biology (alledgedly the source of science in medicine). But it has been objected, by "normativists", that the concepts necessarily refer to values, so that they cannot be defined in ordinary scientific terms. Medical science, if such a thing exists, is linked with practice and that is where values enter the scene. To put it simplistically, disease is a state which people who consult a physician would *like* to get rid of.

We think that current discussions are confused because philosophers of medicine have failed to make distinctions which are commonplace in the philosophy of science. *Concepts can refer to values in different ways.* So can statements. Consider the statement that health is desirable. It may express a conceptual link (" "health" is defined as ...") or an assertion which presupposes that "health" is defined in a different way. In the latter case, it may express a normative stance, but it can also have a descriptive function (cf. "people do in fact desire health"). Distinctions of this kind are seldom made in the philosophy of medicine.

Philosophers of medicine who are inspired by phenomenological philosophy would characterize problems with health and disease in a somewhat different manner. They would argue that the "lifeworld" is the primary locus of health and disease. The concepts used by modern medicine signify an abstraction from experienced reality which, in many cases, is a poor basis for medical treatment. Phenomenologists would agree with normativists that values play a role, but they are mostly concerned with more general issues. They aim at the development of a method which overcomes the limitations of abstraction, the tool of science. Specifically, most phenomenologists think that the "mind-body problem" is an artefact generated by science. In primary experience, there is no distinction of mind and body, of subject and object.

Philosophical *analysis* and phenomenology obviously look at health and disease from rather different angles. Which approach would be the better one? The question is inappropriate. If one has the purpose of clearing up confused discussions, one will benefit from analysis. If the purpose is to infuse the medical image of man with human purpose and meaning, phenomenology is a better option than conceptual analysis.

At this point, we would like to show our hand. We emphatically endorse one form of relativism. General theories, be it in philosophy or in science, are never adequate *simpliciter*. The notion of adequacy makes sense only if it is associated with *criteria* and *purposes* in a particular *context*. This obviously limits possibilities for developing grand, integrative theories. As far as we are concerned, analysing the limitations of grand theories is one of the most important tasks of philosophy.

Attempts to develop such theories are common indeed. Normativists would like to cover empirical and normative aspects of health and disease by a single theoretical framework. Phenomenologists want to develop a view of man which improves on the theories of various fields of science. Psychosomatic medicine continually tries to amalgamate biology and psychology within medicine. And so forth.

We are very critical of such developments. One example, which will play an important role throughout this book, will illustrate how we would like to approach them. Consider the following line of reasoning, which is common in medicine and the philosophy of medicine. "Biological factors are implicated in the etiology of most diseases. But psychosocial factors play a role as well. Adequate theories of medicine will have to deal with relations between the two kinds of factor. Specifically, one could try to understand relations between, say, life events and cardiovascular disease in terms of a theory of stress."

We think that integrative theories which have been developed along such lines are often pseudo-integrations. For example, "stress" has been defined, in the context of "unified" theories, in terms of psychosocial stimuli *and* physiological responses. This hampers the study of empirical relations between such stimuli and responses since relations cannot be conceptual and empirical at the same time. So the connections between psychological and biological theories remain unclear in this case.

The upshot of what we have said so far is that medicine has little *coherence*. Coherence has two aspects. An entity has coherence only if its boundaries are well-defined, i.e. if the *demarcation* between the entity and its surroundings is clear, and if it has *internal* coherence. Although we will not use the terms very often, demarcation and internal coherence are recurrent themes in the book. They will be considered in relation to medicine as a whole, and in relation to sub-disciplines within its domain.

3. THE BOOK

Philosophy books come in kinds. Some contain a single protacted argument which is best assimilated in one stroke. Others present a complicated netwerk of arguments which one will get to know by bits and pieces only. Ours is in between. It is meant to serve two different purposes, so there is more than one way to deal with it.

Firstly, we want to provide information for a general readership among university graduates and postgraduates. Much has already been written on the subjects indicated by chapter headings, but we have tried not to duplicate extant views. We have therefore disregarded medical ethics, the area which gets most attention in the philosophy of medicine.

Current approaches in the philosophy of medicine are often one-sided, unavoidably so in view of the explosion of information due to the growth of research. We have tried to compensate for one-sidedness by behaving as generalists. That is a dangerous option in our time as one's work will easily become superficial. If there is any depth in the book, differences in our background have helped us to reach it; WJS worked in biology and philosophy, PJT started in pathology and later worked in medical education and in "metamedicine".

Secondly, the book is meant to have a function in university courses at the graduate level. For example, students can use it as a basis for training in philosophical analysis applied to medicine; many examples of analysis are found throughout the book. Likewise, the materials presented can be used to evaluate competing views on medicine and allied disciplines.

We do hope that the book will help students to discover and explore new problems.

Acknowledgements. We are indebted to the following colleagues for helpful comments on parts of the book: Paul Doucet, Bas Jongeling, Peter Kirschenmann, Philip Kitcher and Ed Thieme. If the book has any quality, a substantial share of the praise should go to our colleagues Peter Sloep and Bart Voorzanger, who uncovered many shortcomings in successive drafts preceding the final version. Technical assistence was kindly provided by Peter Blaauw, Ted Doove, Thea Laan, Eduard van Staeyen and Ria Stokman.

CHAPTER II. CULTURAL INFUSIONS IN THE PHILOSOPHY OF MEDICINE

1. INTRODUCTION

Consensus is a fairly common phenomenon in science. In philosophy it is not. There is disagreement even over the nature of philosophy.

Philosophy of science is an exception. It is more homogeneous than other branches of philosophy because its subject matter, science, is reasonably well-defined. Should there be homogeneity in the philosophy of medicine as well? One would expect it since medicine contains so much science. But not so. There are few generally endorsed basic principles in the philosophy of medicine. Perhaps this is explained by the impact of medical practice, which is deeply embedded in local cultural traditions.

Throughout this book, we will repeatedly argue that philosophy of medicine (like medicine itself) has much to learn from the philosophy of science, especially methodology. In the present chapter the emphasis is different. We will portray some influential philosophical views of medicine against the background of their cultural setting. Two examples will be analysed in detail, American "progressivism" (our term) as defended by Pellegrino and Thomasma (1981), and anthropological medicine, which developed in continental Europe a few decades ago.

Would it be possible to develop any transculturally valid perspective on medicine? We like to think so. McKeown's (1984) work, which we also analyse, seems to offer opportunities for the development of such a perspective.

2. AN AMERICAN ONTOLOGY

2.1. General survey

Pellegrino and Thomasma are among the leading representatives of USA philosophy of medicine, and the book published by them in 1981 is well-known. Its subject matter covers a lot of territory, and the same holds for its philosophical approach. The authors are avowedly eclectic, they mix philosophies as different as European phenomenology and Anglo-Saxon philosophy of science. Because their view is rather complex, we will give a general survey before commenting on specific theses defended by them.

Pellegrino and Thomasma start from the question of how it is possible to apply biological and pathological knowledge to medicine practised on individuals (p. 42). They regard practice as the core of medicine, and cure as the heart of practice. "More specifically, how can medicine apply theory to a concrete, singular, individual body and produce a cure? A "cure" is a successful medical event. What are the conditions of its possibility?" (p. 83).

This starting point, which recurs in many places (e.g. p. 50, p. 121), is presented as a matter of choice without much justification. Pellegrino and Thomasma admit that "personal concerns", besides "cultural concerns" determine the selection of "what is important" in medicine (p. 49). They decide to ignore other matters which are less important or uninteresting from this point of view.

The authors are aware of the limitations of their starting point, and they concede that one should not restrict the aims of medicine to "curing" and thus disregard "caring" and prevention. Nor should one disregard nature's healing capacity. Moreover, "medicine aims at a restoration which can only be asymtotic to former well-being" (p. 62, p. 63). Yet cure functions as a cornerstone for their philosophy. Culture determines the choice of philosophical method and in the case of medicine this should lead us to the development of a practice-oriented "ontology" centering on the following basic question. "Cures and healing take place. They are experienced. How is that possible?" (p. 57).

In effecting cures, medicine draws on general scientific knowledge of life processes and disease mechanisms. When discussing the value of this kind of knowledge (pp. 108-113), the authors seem to endorse a positive, realistic view of medical science. Thus they argue that common character-istics of human bodies are rightly objectified in medicine's world of symbols. Cures would be impossible if scientific knowledge and reality would not match here. The authors also argue, however, that human beings are unique in the sense that they defy total objectification. Medicine therefore cannot rest content with common science. And it will need much more than science in the way of knowledge.

To understand medicine one must first and foremost concentrate on the clinical event (interaction between physician and patient), the core of medical practice (pp. 62-64, p. 143).

In the clinical setting, medicine effects cures by working in, through and with the body (p. 73, p. 106). A proper understanding of what this means is possible only if various connotations of the word "body" are distinguished. For this purpose, the authors introduce three special terms, "living body", "lived body" and "lived self". This terminology, which is inspired by European phenomenology and existential philosophy, should help one to avoid Cartesian mind-body dualism. We will not fully quote Pellegrino and Thomasma's explications (which are involved) but only try to capture the general idea.

"Living body" refers to the physical organism, interacting with its surroundings in the struggle for survival, in short, as the body dealt with by biochemistry, physiology and other branches of biology, a body invested by a wisdom of its own on which the capacity for survival is built (p. 74, p. 107, p. 111). The living body is held to be fundamental for the possibility of cures: "... the possibility of a cure must ultimately be grounded in a unique organization of matter which we have called the living body" (p. 111).

The term "lived body" refers to the experience of being a body, the subjective side of the human organism tied to a shared intersubjective reality of language, values and society in general (p. 73, p. 74, p. 107). This is the level of everyday reality, which medicine must take into account as it aims at cures. Scientific explanation and manipulation of disease mecha-nisms will not help one much unless they affect "the lived body with all its values" (p. 114). And this requires physician and patient to interact in terms of the everyday world with its intersubjective significance. "Because the cure must be satisfactory and rehabilitative, it must be grounded in a cultural and individual value system created by lived bodies" (p. 114). Both medical practice and medical science are thereby imbued with values.

"Lived self" denotes the historical interpretation of one's experience as a lived body (p. 73). This interpretation, of course, plays a part in the

definition of health and disease (p. 75) and therefore affects clinical interaction (p. 113, p. 179).

The historical (biographical) aspect of the human body links the
"ontology of the body" to a philosophical anthropology (pp. 116-118) in
which personal identity is described as a "symbolization" of the body. "In
this symbolization, medicine presupposes the perfectibility of man, a condition it shares with ethics" (p. 118).

2.2. What place for science?

Pellegrino and Thomasma's views are *progressivistic*. They allot a powerful, constructive role to medicine in our culture.

We want to put progressivism in a proper perspective. Before doing
this, however, we will consider the underlying fabric of philosophical
theses which the authors defend. It is a tight fabric which is not easily
analysed. For ease of exposition, we will here put one theme, the role of
science, at the centre.

We have extracted from the authors' text various theses involving
science which call for criticism .

*Thesis 1. To understand medicine, one must concentrate on medical
practice rather than on medical science.* Remember that medicine's power
to cure is all-important in Pellegrino and Thomasma's philosophy. Curative
intent, for them, is the predominant feature of clinical interaction between
physician and patient, and clinical interaction seems to be the heart of
medical practice. It is not surprising therefore that practice is emphasized.
The following quotation illustrates this.

After choosing what is important about medicine as a starting point, successive
descriptions are necessary about this reality, the doctor-patient relationship. In particular,
the descriptions focus on the formal conditions necessary for its effectiveness. However,
the effectiveness of medicine, either cure or healing, is a *praxis*. Therefore, the method of
philosophy must begin in practice and return thereto for a test of its meaning. In other
words, a philosophy of medicine must be an ontology of practice, a search for meaning
in the practice of medicine, and specific applications of the results of this search (p. 50).

Common (philosophy of) science will not help one here because "it neglects
the richer reality of medical practice" (pp. 50-51). Medical science may be
helpful, but it "neglects the nonmeasurable factors brought to the clinical
relationship by the patient, physician, and the cultural environment" (p.
51). Philosophy of medicine must embrace such factors. How can it do that?

One must concentrate on the "lifeworld" (Husserl's *Lebenswelt*) "as a condition of theory and ideas" (ibid.). That is, "theory must be derived from the lifeworld as its condition of possibility" (ibid.).

The program which Pellegrino and Thomasma sketch presupposes that the lifeworld is a good basis for the development of philosophical theories. The authors have a lot of confidence in the lifeworld. "While mistakes about reality can be made in the lifeworld, they are less apt to occur than in the realm of theory" (p. 54). Indeed, an ontology of practice "entrusts to the lifeworld the task of correcting theoretical postulates" (p. 55). The authors do not offer much in the way of arguments. Perhaps they regard the matter as self-evident because "science, institutions, concepts, organizations, theories, laws, and philosophy itself all depend upon a logically prior world of experience" (p. 54).

Philosophers of science nowadays will not accept this view. One could as well argue that lifeworld experiences are shaped by "logically prior" theories. The authors seem to be aware of this. In discussing experiential judgements in the clinical setting, they concur with modern philosophy of science when they admit that unprejudiced observation is a myth (p. 67). Unfortunately, *this* philosophy is hardly compatible with the kind of phenomenology they seem to defend.

Could the lifeworld give one criteria for correcting science? Or should one use science to correct erroneous views which populate the lifeworld? Such questions are not as easily answered as Pellegrino and Thomasma seem to think. We will consider the issue in greater detail in chapters III and V.

Thesis 2. Medical science cannot capture uniqueness. Any scientific approach of health and disease will involve "objectification".

What is objectified ... is *not* unique. To objectify oneself as a lived self or person, or to objectify disease categories and etiological agents, is to enter the world of common characteristics. Uniqueness lies instead in the historical, space-time configuration of the "capacity to objectify but not totally" - that is, the lived body. ...

Because the world of common objects and the objectified lived self are the results of a capacity to objectify on the part of lived bodies which cannot be objectified, scientific explanation of disease is not the only, or even the major, task of medicine (p. 109).

[Medicine, in dealing with individual patients, aims at *specificity* because it is action oriented. As its endeavours become more specific, uniqueness will play a more prominent role.] The implication of specificity leads to the unique methodology of medicine. ... Physical and laboratory diagnosis is a process of narrowing the range of possibility by comparing a specific body with normal values. The process is governed by the nature of the interaction, a requirement to reach a decision on behalf of the body in need. Hence the method of medicine is not the same as that of the other physical sciences, which have no requirement to act on behalf of a body in need. ... ...clinical judgements differ from those in pure science in that they are governed by individualization. Clinical judgements must be a complex process of perceiving individual uniqueness in the midst of common objectivities (p. 110).

Individualization, in turn, depends upon the physician's discovery and utilization of the wisdom of the body. The response of each patient to therapy cannot be predicted scientifically (p. 111).

Notice that Pellegrino and Thomasma are here amalgamating at least *two* different theses. There are limits to the power of medical science because it cannot deal with (i) uniqueness and (ii) subjectivity. Let us begin with subjectivity. Is it impossible for science to deal with subjectivity? The answer depends on how one uses "subjectivity". Perhaps the pain you feel, *qua* "subjective", inner experience, cannot be captured by science. But we can surely talk with you about the "subjective" state (pain) you are in. And there is no *a priori* reason why we could not go at it in a scientific way. We do not want to elaborate the theme of science and subjectivity, which is no-toriously tricky. Pellegrino and Thomasma obscure the issue by associating subjectivity with uniqueness and specificity. Many of their subsequent arguments concerning limitations of science are couched in terms of uniqueness. However, uniqueness as such need not interfere with a scientific approach. When a particular window breaks by the impact of a stone one can describe what happens as a unique event which neatly obeys the laws of physics.

True, Pellegrino and Thomasma rightly notice that uniqueness in medicine often calls for descriptions in terms of many factors. This will hamper scientific explanation and prediction because one will be plagued by *complexity* (rather than uniqueness!), and by lack of information. As one adds more factors in considering a particular phenomenon there will of course be trouble. Beyond a certain point science may cease to be helpful. In medical practice the problem is a real one. In this respect Pellegrino and Thomasma are right, science and methodology have their limitations. But is it possible to get rid of *these* limitations by attending to the lifeworld? The question seems to remain unanswered. In a different part of the book (chapter 6: "The anatomy of clinical judgements") the authors suggest that there is no answer now. They mention philosophers like Polanyi, who have stressed that there is something like "tacit knowledge" besides scientific knowledge as ordinarily characterized (p.141). *Experience* may help us where the limits of a scientific approach are reached. Unfortunately, there is no methodology of "experience". Just now we will not consider the "experience problem" in greater detail. It will get more attention in chapters III and V.

The analysis given so far does not cover all the items which Pellegrino and Thomasma bring to bear on the uniqueness problem. The additional issue of values affecting medical treatment must be considered as well. Thesis 5 deals with this issue. Before discussing it, we will focus on two other theses.

Thesis 3. Medical science has failed to solve the mind-body problem. Thesis 4. Medical science (like medicine as a whole) should concentrate on the body. These theses are two sides of the same coin. Pellegrino and Thomasma reject Cartesian dualism which still haunts science. Their ontology of the body is meant as an alternative.

We agree with the first thesis, but this does not make us accept the second one as a solution. As long as mind-body puzzles are embedded in science and philosophy, the mere introduction of a new terminology will not be helpful. We do not think that Pellegrino and Thomasma's distinctions (lived body, living body, lived self; see previous section) amount to much more than that. They use expressions like "bodies ... capable of creating scientific formulations" (p. 108), "body in need of help" (many places). "Body" here clearly stands for "person". Now it is not easy to formulate theories about persons (there is indeed a mind-body problem), and changing the terminology will not make the problem go away. The term "living body" is indeed *defined* in terms of the *experience* of being a *body*. It is easy to give this a dualistic interpretation. That, of course, is not the authors' intention, but we are not offered enough philosophical tools to understand how dualism is to be avoided here.

The empasis on the body is intended as a criterion of *demarcation* to distinguish the area of medicine from other areas. We are very skeptical of the criterion because "body" is used in an over-inclusive sense.

Thesis 5. Medicine cannot be value-free. As we saw before (cf. thesis 2), Pellegrino and Thomasma argue that medical science has limitations vis-a-vis medical practice because practice cannot disregard the *subjective* side of *individuals*, who are *unique*. According to them, this implies that *values* unavoidably belong to medicine, even medical theory. They regard medical ethics as an intrinsic part of medicine.

... Medical ethics is not a body of ethical theory applied to medical transactions but an intrinsic part of medicine itself. Its argumentation is, therefore, based on values perceived within the medical relationship and not external or relative to it (p. 174).

The phrase "based on values perceived" is intriguing. *Why* are values of the persons concerned a sufficient basis for ethics? We think that Pellegrino and Thomasma have not really addressed this question in their book.

"We have seen that the primary meaning of health is grounded in a condition of the living body, and that the norm for medical ethics is also grounded in that living body" (p. 183). How is the grounding done? We hardly get an answer. It almost seems as if Pellegrino and Thomasma feel able to derive values from factual matters, a process which many philosophers brandish as a *naturalistic fallacy*. On the same page, however, the authors state that they have not argued "according to the lines of the naturalistic fallacy that science itself, or even medicine itself, can determine values which are right or wrong". How then are we to understand the following passages?

What we propose is a mutually binding set of obligations, predicated upon a special kind of human interaction and deriving its morality from the empirical realities in the relationship which specify it among human relationships (p. 218).

We proposed to found professional ethics in the *fact* of illness and the *act* of *profession* (p. 232).

All rights of any substance are based on some common fact specific to human existence (p. 237).

This *sounds* like naturalism, but perhaps the authors do not precisely say what they mean.

The relations between facts and values confront one with yet another *demarcation* problem. Pellegrino and Thomasma, in our view, have not contributed to its solution because they do not make appropriate distinctions. We will introduce some distinctions in the next section which deals with general values defended by Pellegrino and Thomasma. A specific variant of the issue (Are the concepts of health and disease value-laden?) will be considered in more detail in chapter IV.

2.3. The riddle of "ought" and "is"

It is good to be honest. Medicine aims at cures. Physicians ought to listen to their patients.

These statements all involve values and/or norms. (The distinction of values and norms is important, but for the present purpose it does not matter much.) Should one conclude that the statements are unempirical, or that they do more than *merely* describe facts? Ethical naturalists would deny this. They would aim to define "good" in terms of "natural" properties. And they would try to translate sentences with "ought" into factual statements with "is". Naturalism is not popular today, though there are signs of a revival (see e.g. Richards, 1986, and the discussion following his article). The intended translations would anyhow be far from easy.

Let us assume that the statements of science are empirical, or at least that they are meant to be empirical. Would that imply that science is value-free? That depends. If naturalism would be acceptable, science would not need to be value-free in order to be empirical. Now it is tempting to suppose that the rejection of naturalism would entail that science is value-free (to the extent that it is empirical). But the temptation should be resisted until we have a good grasp of terms like "value-free".

Under non-naturalism, "It is good to be honest" is not a promising candidate in the search for empirical statements. "Good" is here an obvious stumbling-block. What about "honest"? Could "honesty" be a concept of science? Here one needs to be subtle. How is "honesty" to be defined? *We* associate the concept with things like telling the truth. If that is reasonable, "most people are honest" (as opposed to "It is good to be honest") is not a value-statement. It is factual (descriptive, empirical) because facts determine whether it is true or false. Now you may disagree with us, and opt for a definition which makes "honesty" value-laden. Would that make "most people are honest" a value-statement? In one sense it would indeed. The statement would involve values in the sense of *mentioning* them. Unlike "It is good to be honest", however, it would not *express* values. *Values can be mentioned in a descriptive way.* If one forgets about that, discussions about science and values will easily become a morass.

Like Pellegrino and Thomasma, we will assume that naturalism must be rejected. But we do not conclude that "pure" science cannot be concerned with values. It clearly can, in many ways. What about the other statements we began with? "Medicine aims at cures" clearly *involves* values. But it can be interpreted as a factual statement. "Physicians ought to listen ..." is different. It involves values in such a way that we had better regard it as an unempirical statement.

Our account of facts and values is rather simplistic (for additional comments, see chapter IV), but it will serve to get Pellegrino and Thomasma's views in perspective. To avoid confusion, we will reserve the term "normative statement" for statements which *express* norms or values. The term "factual (empirical) statement" will be used in the normal broad sense which covers statements descriptively mentioning values.

Here are some examples of statements in the two categories.

Factual statements. "A majority of medical interventions based on modern technology does not lead to effective cure." "Medical technology is a source of illness besides cure." "A heavy emphasis on cure, through side-effects of therapeutic efforts, tends to decrease general well-being in chronically ill people."

Normative statements. "Practical medicine should aim at the development of techniques serving effective cure." "Cure should have the central place in philosophical reflection on medicine."

As it happens, we think that the above factual statements may well be true. We are also tempted to reject the normative statements because they could lead to unpalatable consequences.

Our view of the matter will be developed in later sections. The point of the examples is merely to reveal the bias in Pellegrino and Thomasma's arguments. Their development of what they call "ontology of practice", in our opinion, conflates factual and normative matters that had better be kept apart.

Pellegrino and Thomasma hardly consider the possibility that medicine may have destructive effects. The (realistic) aim of medicine as they view it is "not entirely pragmatic or utilitarian, but is actually a good, a healing, a perfection of human nature" (p. 180). And while the perfectibility of man is qualified by the acknowledgement that sorrow and death are part of the tragic condition of man, medicine is held at least to make the condition bearable (p. 118). How can they reach this positive conclusion? We think that it is based on a fallacious conflation of factual and normative statements. There is a surreptitious change from "ought" to "is" which suggests that medicine is as they want it to be. By way of an illustration, consider the following examples.

The authors repeatedly argue that their philosophical views are "grounded" in medical practice. General statements to this effect, however, are inherently ambiguous. "Medical practice", in one interpretation, could refer to factual matters, i.e. to situations which are typical of actual medicine. Or it could stand for desirable practice, a normative matter. Any realistic view of the facts of practice, in our opinion, will show that Pellegrino and Thomasma should have the latter interpretation in mind. But they do not make this explicit. Precisely this enables them to keep negative aspects of medical practice (a factual matter) outside their perspective, although their formulations suggest that the facts prove them right.

Consider the following, more complicated statement reflecting commitments in medical ethics, which is open to various interpretations. "Neglect of the norms in the practice of medicine is a neglect of the nature of medicine itself" (p. 191). "Norms" could refer to views which are actually endorsed by the medical profession. Or the term could refer to an ideal body of ethics beyond commonly accepted views. Likewise, "nature of medicine" is ambiguous. It could stand for the character of medicine as it exists today (a factual matter). More probably the authors want to indicate what medicine should be like (a normative matter).

Similar comments apply to more general views professed by Pellegrino and Thomasma. "Civilization is ultimately an organized effort of men to secure the good ends of human life" (p. 171). Taken as a descriptive statement, this reflects a rather optimistic view of civilization. A pessimist, with equal justification, could describe civilization as a random interaction of men striving at self-aggrandizement and mutual destruction. More probably, however, the authors are rendering a normative view. By giving it the form of a factual statement, they are shutting their eyes to discrepancies between ideals and reality.

The statement about civilization is made by the authors when they introduce medical ethics (chapter 8). "Medicine" in this context gets the above role of "civilization". It is portrayed as a beneficial enterprise with effective cure at the centre. Pellegrino and Thomasma do not make a clear difference, however, between saying that medicine *is* beneficial and saying that it *ought* to be beneficial. Similarly, *is* cure at the centre of medicine or *should* it be at the centre? The short-circuit between is and ought lurks here as well.

As a result of this approach, the book hides from its readers both bad medicine and medical interventions with undesirable effects. This American "ontology" thus suggests, not only that evil always loses in the end, but even that it does not really exist!

2.4. Medical practice revisited

The over-intimate links which Pellegrino and Thomasma forge between "ought" and "is" fits their general approach of philosophy and medicine. In the preface of their book (pp. VII-X) it is argued that we need an integrative and explanatory philosophy, which should indicate and clarify the goals of medicine in terms of a philosophical anthropology defining the nature of human beings. At the same time, the authors state that "the theory discussed in this volume stems from our understanding and description of the realities of medical practice" (p. VII) and that "medicine can offer the verifiable, empirically sound phenomena upon which contemporary philosophical anthropology might be based". Then, in a prologue (pp. 3-5), they describe their enterprise as "the way we construe medicine and justify its needs and purposes". They have thus come full circle: medicine needs the guidance of a philosophy which the authors will derive from the empirical reality of medicine. At the same time, the ensuing philosophy is meant to serve the interpretation and justification of medicine.

So far, we have merely shown that Pellegrino and Thomasma's views are biased for theoretical reasons. To strengthen our case we will confront it with the practice of health care.

Medicine in the U.S.A. has been subjected to severe criticism in the last two decades. Educationalists and sociologists led the attack. They were joined later by economists and philosophers. But Pellegrino and Thomasma almost totally disregard contemporary critics such as Carlson (1975), Freidson (1976), Illich (1975), Kass (1975), Kosa and Zola (1975), and Navarro (1980). Only Illich and Carlson are briefly mentioned by them.

Some years ago Michaelson (1981) reviewed a number of critical accounts of American health care in a paper entitled "The coming medical war". The conflict to which he referred is generated by a discrepancy which is well documented in the books he was reviewing. On the one hand, "the medical profession in the United States is extraordinarily rich and its medical scientists are dominant in international research". On the other hand, "citizens of the United States ... are among the least healthy people in the industrialized world". The author analysed various explanations of this discrepancy and proposals for its elimination. His final conclusion was that "the war for a decent change to live a healthy life has started too late, but there are signs, at least, that it has started".

That war has raged indeed for several years without basically affecting health care. One should not have expected anything else in view of the close relations between the general cultural climate, and professional attitudes and organizational characteristics of health care. In medicine, as in other fields, the U.S.A. lead the way of scientific and technological progress. In health care, as in other social services, they are lagging behind most other western industrialized countries in attempts to cover the needs of the population equitably. With a surplus of hospital beds and physicians, health care costs have risen to over 10% of the GNP (1983), while some 28 million people are not yet covered by federal or private insurance programs. These data were recently quoted by the president of Harvard University (Bok, 1984) to demonstrate the need for changes in medicine and in the health care system. One factor which impedes such changes is the resistence to change in medical education. "In all professions, formal eduction is shaped to fit the prevailing sense of how practioners go about resolving the characteristic problems of their calling" (p. 35). In order to break away from this "prevailing sense" and thus to influence the way physicians practice their craft, Harvard's medical school in 1983 started a pilot program in curriculum reform. The program aims at changeing knowledge and skills of young doctors *and* their attitudes towards patients and society in general. Attitudinal items like "recognition of how financial aspects of practice affect self, patients and society" (Bok, l.c., p. 43) are indeed at variance with dominant medical practice in the U.S.

It should be realized that the "prevailing sense" in Western Europe is somewhat different. A study of four university hospitals in Western Europe (in Belgium, Great Britain, Holland, and Western Germany) and one in the U.S.A. showed that the U.S. hospital, compared to the European ones, is more expensive, has more physicians and other employees, more advanced technology, and more intensive care beds (Schroeder, 1984a). All this is parallelled by a higher level of research activity, and a more aggressive clinical approach to severe and to potentially terminal illness. The latter features, however, were also observed in the West German hospital which, like the American one, harboured more severely ill patients than the other European ones.

These conclusions, although based on the study of five hospitals only, probably reflect general trends. High technology like CT scanning and ultrasound, and advanced treatments like coronary artery surgery, kidney transplantation and neonatal intensive care, are more common in the U.S. than in Western Europe (cf. references in Schroeder, 1984a). Schroeder noticed that emphasis on high technology is stimulated by patients or their relatives demanding heroic treatment for advanced illness. "The assertive behavior of U.S. (and German) patients must be seen as part of a larger pattern including high use of medical technology, aggressive treatment of terminal illness, and a higher rate of medical malpractice litigation" (p. 245).

A subsequent analysis concerning the organization of the medical profession in the same countries (Schroeder, 1984b) also indicates that the medical "climate" in the U.S. is different from that in Western Europe, even when one allows for differences between European countries like West Germany and Great Britain. The U.S., which has by far the highest percentage of medical specialists, "stands alone in the extent to which trained specialists - internists, pediatricians, and gynaecologists - serve as primary physicians" (p. 375). In this study, the U.S. also was the only country where primary care physicians continued caring for their patients after hospitalization. The public in the U.S. generally gains access to expensive technology and highly trained specialists more frequently and more easily than the public in Europe.

The differences in economy, technology and facilities for therapy noted above, in our opinion, are a symptom of general cultural differences. In the U.S. both the profession and the general public have more trust and higher expectations viv-a-vis modern medicine. Their reliance on the possibility of cure is related to an emphasis on the perfectibility of man as viewed by Pellegrino and Thomasma. Symptomatic is a definition of "basic health care" given by an American philosopher, Ozar (1983), who postulates that its main concern is the availability of emergency procedures in life threatening situations. The need of highly trained and technically well-equipped personnel for emergency cases, is here exemplary for the most basic level in the hierarchy of human needs!

Small wonder then that in Pellegrino and Thomasma's book contemporary sceptics and critics of medicine are practically ignored. Medicine, with all its "impressive and pervasive modern capabilities" (p. 3) is taken at face value and its history is viewed as a persistent and hopeful uphill struggle towards ever increasing knowledge and skills. Philosophy is called in because medicine is plagued by the complexities of its role in modern society: there is a need for new integration, explanation and justification because these are hard days. But they are wonderful days too, full of promise, and medicine - philosophically beautified - will help to realize the best of possible worlds.

Note added after the completion of the manuscript. Various Dutch colleagues of ours have recently discussed Pellegrino and Thomasma's book in a special issue of *Theoretical Medicine* (vol. 8, 2, 1987).

The contributions in *TM* represent various schools in the philosophy of medicine. For example, Hertogh's (1987) view belongs to a philosophical tradition established by Canguilhem, Tiemersma (1987) draws on Levinas' philosophy, and Richters and Bonsel (1987) develop views on the basis of Habermas' hermeneutics.

In the preface to the special issue, Ten Have, Bergsma and Broekman (1987) comment that the papers "demonstrate something which is typical of European philosophy (of medicine): The existence of fundamentally incompatible systems and schools of thought. ... The authors in this issue do not really dispute Pellegrino and Thomasma's views, rather, their views have induced and provoked the authors into giving us an exposition of their own views on medical matters" (p. 100).

In the face of disagreement among schools, Ten Have, Bergsma and Broekman opt for tolerance. "The recurrent denunciation of other people's ideas and theories is the *real* scandal of philosophy, not the fact that so many different theories and philosophical schools exist. Diversity and plurality are the hallmarks of philosophy" (p. 100).

We agree up to a point. Disputes among philosophical schools should not degenerate into an exchange of mere destructive criticism. But we would add that tolerance can go too far. For one thing, there are elementary criteria for evaluating philosophical work which do not belong to any particular school. Clarity of expression is such a criterion.

As far as we are concerned, Pellegrino and Thomasma's book does not meet this criterion in a sufficient degree. Our comments do amount to a "denunciation" of their "ideas and theories". We have argued that the book is not sufficiently clear, and that central arguments it contains seem to be fallacious. Moreover, the book is in an important sense beside the point (beside real medicine, that is).

Verwey (1987), in the *TM* issue, is the only author who comes close to this verdict. His approach resembles ours, but he has emphasized other issues than we have.

The replies given by Thomasma and Pellegrino (1987) do not make us feel that our comments need to be revised.

3. THE CASE OF ANTHROPOLOGICAL MEDICINE

3.1. European sources

The optimistic philosophy of medicine defended by Pellegrino and Thomasma is actually an ambitious philosophy of man and his culture. The theory is extremely one-sided. Pellegrino and Thomasma uncritically accept modern medicine as a beneficial force, and they almost disregard its critics. They also fail to test their philosophy against existing alternatives which merit attention.

We want to consider one alternative with a venerable tradition. It may be characterized as the personalistic/phenomenological anthropology of the first half of this century. Merleau-Ponty is among its leading theoreticians. Pellegrino and Thomasma do consider Merleau-Ponty's philosophy of the body, but they disregard his denunciation of the treatment which the lived body (*le corps propre*) suffers from medical physiology (and, one should add, psychology).

Merleau-Ponty is but one of the European philosophers who between 1920 and 1960 influenced the philosophy of medicine. In the tradition from Scheler to Sartre and from Husserl to Plessner, these philosophers developed diverse approaches that do not really constitute one movement. Yet these approaches eventually combined to produce a characteristic view of medicine. It is a view which strongly contrasts with Pellegrino and Thomasma's philosophy, their reference to European authors notwithstanding. "Anthropological medicine" is commonly used as a label for this view, which developed in the decades between 1920 and 1960. (Its ancestry actually goes back to the 18th century.) After some 20 years of relative obscurity it now seems to interest philosophers of medicine as well as practitioners anew, in the USA and in continental Europe.

Anthropological medicine is not a unitary movement, but its divergent ideas concerning man and medicine have various basic notions and practical applications in common. It originated in the twenties in Germany and later spread mainly to France, Belgium, the Netherlands and Switzerland. Inspired by various philosophers, authors in the field of medicine tried to expand the image of man prevailing in medical (including psychiatric) practice with a phenomenological or personalistic anthropology. We will not try to describe the similarities and differences among representatives like Von Weizsäcker, Von Gebsattel, Boss and Buytendijk, but only contrast some of the general ideas they share, with the American view discussed in section 2.

The human body or, rather, bodily existence, is a good starting point. The authors mentioned all protest against Cartesian dualism, which they replace by a holistic view of man. According to the anthropological view, bodily existence is embedded in nature, in organic life. At the same time, however, it expresses a wholeness of its own which confers meaning to the body and its relations with nature. Meaning and wholeness obviously have a historical dimension. Every human being continually interacts with the natural environment (internal and external), but episodes of interaction cannot be divorced from past and future biographic patterns.

Some religiously inspired authors elaborate this view of man in terms of creation. For others, relations with a creator are not essential. All representatives of anthropological medicine, however, agree that man must be characterized in terms of historical rather than biological holism. The "whole man" as a person unites past, present and future in a pattern of meaning that transcends the domains of physiology and psychology.

The survey given in the next section will show that this view, despite features shared with Pellegrino and Thomasma's "ontology of the body", leads to quite different conclusions. Critical comments will be given in section 3.3.

3.2. Beyond the body

So far, we have characterized anthropological medicine in vague and abstract terms. Let us consider, more concretely, how it deals with health and disease, diagnosis and treatment. For Von Weizsäcker (1950), episodes of illness are expressions of our fundamental inadequacy (*Unzuläng-lichkeit*) and steps on the path to our ultimate destination. Diseases may be analysed at the level of somatic or psychosomatic mechanisms. Personal life history, however, is invariably involved as an additional level. Medical diagnosis and treatment become meaningful only at this level. Every disease is to some extent incurable in the context of personal life history. It is a fractional death since it always leaves a trace, a scar, a reminder of our fragility, a *memento mori*. In this context, "healing" is not an aim in itself but rather a means by which medicine may help people to follow their path of life. Patients naturally demand cures and physicians do have the task of promoting cure. But the restoration of health cannot be the ultimate goal. Patient and physician alike will have to face the partial incurability of all diseases and they should consider the personal, biographical meaning of each illness.

These views have obvious consequences for the practice of medicine. Von Weizsäcker (1950), for example, decidedly condemns persuasive behaviour of physicians which may effect placebo cures. It leads to transactions between physician and patient which he labels as mutual stupefaction (*Verdummung zu zweien*). Where doctor and patient help each other to evade clarity, medicine forgoes the responsibility it has beyond therapy, to help patients fulfil their human destiny (pp. 130-135).

Von Weizsäcker is widely regarded as a typical representative of European anthropological medicine. His ideas, however, are specific in view of his religious motivation. Moreover, his writings concern medicine as it existed in the thirties and fourties. To prevent one-sidedness, we shall also consider a more recent representative of the anthropological trend who does not approach medicine from a religious perspective, the Swiss psychiatrist and philosopher of medicine Medard Boss. In his writings, the influences of phenomenology on anthropological medicine are very clear (Boss, 1971). The image of man developed by natural science, according to him, is not merely a fragmentary reconstruction to be supplemented by medical psychology and sociology, but even a deformity. Medicine needs a new theory, based on an analysis of our being-in-the-world (*Daseins-gemässe Untersuchungsmethode*).

This analysis leads to views on the human body and on health, disease and therapy, which differ fundamentally from those of scientific or even psychosomatic medicine. Boss (1971, p. 323) agrees with Sartre's view that man can be understood only in relation to his world. Our existence always is a relative existence. There is no separate me, but only me at this table, on this street, oriented towards whatever holds my attention. Existence thus transcends the body as an instrument: man's consciousness, though emer-geing from the body, takes shape only in his meaningful relations with the things of his world. Merleau-Ponty elaborated this philosophical theme into a much more comprehensive picture of bodily existence.

Both Sartre and Merleau-Ponty, however, still considered the body as a "residence" for our conscious existence. So they did not really escape the hold of Descartes' dualism. According to Boss, their philosophy is but a first step towards a truly existential view (*eine daseinsgemässe Sicht*) of man. Unlike the French phenomenologists, he wants us to see our physical existence as a continuous process of shifting relations, of ever changing patterns of embodiment. In this process, the involvement of our body will depend on our "world-relations" of the moment. Even if one is sitting in a chair, one's conscious existence may extend to places far away and times far back, until the chair cracks and a finger is hurt. One's embodiment then shrinks into the finger and time contracts to the present which calls for medical action (pp. 462-463).

This approach leads to a phenomenological revision of psycho-somatic medicine. Rather then adding up somatic, psychological and social aspects, it should develop a more integrative view of existence by con-centrating on meaningful relations with the world which we actually experience. A radical re-evaluation of disease, pathogenesis and therapy ensues. Being human consists of being open to the world and its meanings. This condition, *die vernehmende Weltoffenkeit*, changes in disease. One must refrain from conceptualizing diseases either as physical accidents with psychological implications, or as psychosocial disturbances with somatic manifestations. Both approaches fail to envisage what it is to exist as a human being, in illness or in health.

3.3. The avenue of morality

Boss, like other representatives of the anthropological approach, is not very positive about orthodox medicine. As medicine emphasizes somatic pathology it cannot really come to grips with any episode of illness. Pathology is but one aspect of what happens to us when we are ill. Medical therapy therefore cannot heal patients. It merely provides the preliminaries for reconstructing a free and open "being in the world". If a patient regains his health after a purely surgical or pharmacological intervention, he was essentially healthy enough to regain his full existence. In many cases, somatically oriented medicine does obstruct the process of healing by overlooking what the patient needs. After all, disease resides not in the body but rather in man's relation to his world (Boss, 1971, pp. 554-570).

This is especially clear in diseases which are dominant in modern industrial society. Boss, following Jores (1966), calls them specifically human diseases. Thus "essential" hypertension, chronic gastro-intestinal disturbances, asthmatic disease and other ills, reveal pathological social relations, or rather social deprivations. Here medicine, by concentrating on the physical pathology of individuals, condones fundamentally pathogenic mechanisms in our society. Science, technology and industry are enclosing human beings, cutting off all opportunity for living in a world of open and meaningful relations (pp. 435-436).

For Boss, in short, medicine too often serves the wrong interests by merely returning the patient to the social roles which caused his illness in the first place. Physicians who fight for their "freedom of medical practice" do not realize that medicine as they practice it hardly serves the freedom of patients as human beings (pp. 567-569).

Von Weizsäcker's and Boss' variants of anthropological medicine reflect a view of culture which asks people to realize their own destiny and to follow a transcendental vocation. This view is coloured by pessimism. In this respect, anthropological medicine is an antipode of the American view we analysed. One should realize that the ideology of progress has never caught on in continental Europe as it did in the USA. Although there have been fluctuations in pessimism and optimism in Europe, the second world war was never regarded there as a "war to end all wars". Neither will the possibility of a future war be considered in this light.

The present continental sense of doom expressed in literature, arts and public opinion polls, is not a new development, but rather a reaffirmation of an old latent trend. The anthropological approach of medicine belongs to this trend. It views illness as one aspect of a more inclusive human tragedy. So it is not surprising that it asks medicine to move beyond its own borders. If this appeal is granted, medicine will get the role of helping people to fulfil their destiny, either by realizing their earthly mortality as God's creatures, or by living a free existence in open relations with a meaningful though oppressive world. In either case, medicine will carry a substantial moral message.

Should one be happy with this development? We are critical, though we regard the philosophy of anthropological medicine as more solid and coherent than that of Pellegrino and Thomasma. Let us make some comparisons.

Both philosophies tend to abolish the demarcation between medicine and other domains of culture. Both put the "lifeworld" at the centre of medicine and thereby stress the limitations of a scientific approach. But they do this in very different ways. Pellegrino and Thomasma do not really elaborate a new theory and many of their arguments are fallacious. Anthropological medicine is different. It has developed a real body of "theory" resting on quite a tradition, which cannot be dismissed that easily. So it must be evaluated in other terms.

The "theory" of anthropological medicine is not a theory in the primary sense of the term. It purports to be close to everyday life, and it incorporates elements outside the scope of normal academic science (and much philosophy). What current *discipline* indeed could come to terms with subjects like "the meaning of life"?

It is difficult to *argue* about issues like "the meaning of life". Therefore, we will not attempt to criticize the *"theory"* of anthropological medicine. In fact, we *feel* a lot of agreement! But we do have a different kind of criticism. Anthropological medicine could only effectively treat patients who share its view of life. However, many people have an entirely different view. So anthropological medicine could only be a powerful force for the general public if it would force its philosophy upon patients irrespective of their own affinities. That we cannot accept.

Anthropological medicine is not alone in embedding morality in medicine. Pellegrino and Thomasma's views also seem to result in a moral medicine. They even *say* that ethics (medical ethics at least, presumably much more than that) is part of medicine. Our main objection to their philosophy is that they distort *factual* issues when they take a moral stance (cf. their portrayal of medicine as a beneficial force; here anthropological medicine is much more realistic). This makes them overemphasize the importance of cure in medicine.

Pellegrino and Thomasma's ethics for medical practice is connected with the emphasis on cure and on clinical interaction between physician and patient. Working towards the realization of cure, according to them, is imperative for both physicians and patients. The physician-patient relation is imbalanced because of the patient's vulnerability and the physician's competence and knowledge. This gives the physician a responsibility to bridge the information gap and to enable the patient to join him in decisions about what should be done, while the patient is bound to put his trust in the physician and to respect his professional competence (chapter 9 of the book). Despite their arguments concerning the limitations of medical science (see section 2.2), medical knowledge so gets a crucial role. The central question is what should be done on the basis of this knowledge.

The anthropological physician, on the other hand, is more likely to give precedence to the kind of knowledge which is the privilege of the patient, knowledge about *the patient's world*. What meaning does it have for him? How did it break down? The central question here is not what should be *done* (by the physician) but rather how the patient should restore meaningful relations with his world (with the physician as a sympathetic observer or at most a catalyst).

Both points of view are visibly influenced by the cultural climate. Naturally, any attempt to completely divorce medicine from other domains of culture would be futile, even foolish. But one needs to have *some* demarcation. The two philosophies we discussed obviously give medicine great roles which it is not fit to play.

The moral bias in either view is associated with an *individualistic* approach. Anthropological medicine, to be true, recognizes the impact of the social environment and its pathogenic streaks. (It vastly underrates the impact of the physical environment.) But the remedy it has proposed mostly centres on "improving" people rather than the environment. Individualistic medicine, in our view, is nowadays a bad option. McKeown's analysis of medicine, which is discussed in the next section, shows that it is at variance with realistic strategies to promote health.

4. GETTING THE RECORD STRAIGHT

4.1. The fruits of the past

The two philosophies discussed above are both at odds with medical reality. Proponents of anthropological medicine would readily concede this. They would argue that conventional medicine is rooted in a culture from which man should be delivered since it has estranged him from his true being. Any philosophical commitment to conventional medicine would contribute to the continuation of modern man's estrangement.

The progressivism of Pellegrino and Thomasma, however, explicitly claims medical practice as its starting point. But their view of practice is strongly biased because it emphasizes cure or, more generally, effective treatment. Cures actually represent exceptions rather than a rule, especially in present-day industrial society with its high incidence of chronic, geriatric and psychosomatic diseases.

Has medicine ever been more effective? Popular history suggests that it should get the credit of greatly improved health in the past few centuries. But let us be careful. Medicine has fought the great diseases of the past, and many of them have almost disappeared. Most of the credit, however, should go to better food and hygiene, not curative medicine (McKeown, 1984).

Today, in the western world, medical intervention is often not necessary for cure. Neither is it sufficient. Many common diseases are either self-limiting, or chronic and basically incurable (see e.g. McKeown, 1983). None the less, curative medicine inspired by technology is now a dominant force in our culture. Moreover, western medicine dominates over other diagnostic and therapeutic systems in practically all countries. This is but one symptom of the world-wide expansion of western culture. The penetration of so-called western diseases in non-western societies is another typical symptom (Trowell and Burkitt, 1981).

Since our brand of medicine was developed in western Europe, mainly in the 19th and 20th centuries, the diseases of 19th century Europe were its original area of preoccupation. The modern *western* diseases, which by and large coincide with specifically human diseases *sensu* Jores (see section 3.3) are a stumbling block for western medicine!

What philosophy of medicine should one adopt in the face of these discrepancies? First and foremost, let us refrain from *a priori* moral commitments. Morality divorced from facts is blind. Pellegrino and Thomasma looked at medicine as a moral enterprise. A normative stance made them put *cure* at the heart of medicine without bothering to consider the *facts* about cure. Anthropological medicine, in defending a moralistic view of human destiny, considered what human needs should be like without any extensive inventory of actual needs. Such approaches will not do.

At this point, McKeown's pragmatic analysis provides helpful materials for an alternative approach. Unphilosophical in itself, it unwraps medicine from its cultural context. It provides us with a matter-of-fact perspective on medical theory and practice, which may become a basis for a less biased philosophy of medicine. We will first give an overview of McKeown's ideas, and then comment on specific issues raised by him.

4.2. Towards a realistic view of disease

The present civilization of man emerged after some 300,000 years of adaptation to nomadic life. Our ancestors were hunter-gatherers with a low population density. Subsequent changes in culture involved settlements, the use of domesticated animals (about 10,000 years ago), agricultural and cattle-raising techniques (greatly improved about 300 years ago) and finally industrial development (about 200 years ago). It is reasonable to assume that our somatic make-up did not change as drastically. Many human diseases characterizing our civilization reflect changed habits rather than changes in genetic make-up.

Historical analysis of changed disease patterns does show that genetic changes have at best played a minor role. It also shows that medical interventions concentrating on disease mechanisms were but moderately effective. Changes in the environment and in behaviour (cf. food, hygiene) were much more important. McKeown (1984) presents an impressive amount of data illustrating this.

What about the role of medicine in our day? McKeown gives an assessment for various categories of disease. Firstly, there remain diseases which are prenatally determined by genetical "errors" or intra-uterine influences, some of them requiring our protective culture for persistence. Secondly, mishaps such as accidents and infections are more typically associated with our culture. We promote them by our way of life, but we are often able to combat them as well. Thirdly, there are many diseases associated with our culture which leave us almost powerless. They constitute the most important category. Today's major inflictions reflect our inability to cope with the industrial way of life which now characterizes western culture. Most chronic diseases, diseases of old age and psychosomatic illnesses, belong to this category.

McKeown argues that prevention could lead to a marked decrease of prenatally determined diseases (cf. genetical counselling, techniques for early detection of embryonal deficiences). Accidents, infections, and diseases of malnutrition are increasingly amenable to prevention and cure, although the hazards of drug resistence and other effects of mutations in pathogens or vector organisms will remain. Diseases and illnesses in the last group (inability to cope) are the major source of health problems in affluent societies. They should be preventable to a large degree, but current trends in our culture do not make the realization of prevention easy. Individual therapy will so remain a necessity. For the foreseeable future, this will involve surgical, pharmacological or mechanical adaptation, and palliation and prosthesis for fundamentally irremediable conditions.

It will be clear from this panorama of disease that medicine, in aiming at the promotion of health, should concentrate primarily on the control of social and environmental factors. McKeown is fairly optimistic about human educabilility with respect to such factors. He thinks that a substantial reduction of poverty and of unequal distribution of preventive sanitary facilities should be feasible. His main point is that medicine should shift the focus of its responsibility from cure to care, and from individual therapy to large-scale changes required for more adequate ways of life. In summary, the task of medicine is "to assist us to come safely into the world and comfortably out of it, and during life to protect the well and care for the sick and disabled" (1984, p. 192).

4.3. Biology reinstated

McKeown's view emphatically is a normative one. His intention is to formulate new goals for medicine. The emphasis in his arguments, however, is on how one must deal with factual issues. Orthodox medicine, in the past few centuries, has been dominated by the Cartesian heritage. It concentrated on the mechanics of processes in the body, and almost disregarded environmental and behavioural aspects of disease. McKeown's arguments show that scientific disciplines must be reallocated within medicine and health care. Medicine now draws heavily on biology, rightly so, but it has a one-sided predilection for anatomy and physiology, and allied disciplines. It clearly needs more ecology. We can but agree. Many examples in later chapters will show that distortions of biology are common in medicine.

If one gives the environment a proper place in medicine, one must beware of making it the antipode of genetic factors. Diseases, like any property of organisms, are not determined *either* by innate dispositions *or* by external factors. Any property whatsoever depends on both kinds of factor. Only *differences* in properties between organisms may depend on the environment alone or on genetic factors (but mostly they will depend on both). Those medical authors who consider the environment, unfortunately, do not always make the appropriate distinctions (examples will be given in later chapters). McKeown is not among them. His analysis of genetic and environmental influences on disease (chapter 2) is an example of lucid biology.

It is not easy to disentangle genetic and environmental contributions to any particular disease. McKeown thinks that environmental factors and behaviour have a relatively important role in many diseases. He is rather positive about possibilities for improving health by appropriate measures (example: policies promoting the consumption of healthy food). We agree with the main thrust of his reasoning, but we like to discuss some possible flaws.

McKeown makes population regulation and natural selection important themes. He roughly reasons as follows. "The key to the riddle presented by the health of living things is the relation of fertility to mortality. Both have evolved through natural selection; but they have not evolved in balance, in the sense that numbers born are restricted with regard for the resources of the environment and the numbers that can survive" (p. 5). Thus external factors (cf. food shortages, pathogens, predators) have a role to play in population regulation.

The domestication of plants and animals shows how such factors can be manipulated. More and better food, and removal of other causes of mortality (predators) have made successful exploitation possible. "The methods which have been exploited in plant and animal husbandry are essentially population methods which owe little to understanding of structure and function" (p. 6). There are various reasons for adopting such methods. The most interesting one is that "population methods are far more effective than individual methods" (p. 7). That is a nice lesson for medicine.

McKeown thinks that most diseases are more likely to be controlled by removing causes (at the population level) than by intervening in disease mechanisms. In the following passage, McKeown applies these general ideas to the situation of man.

Like other living things, man has been exposed to rigorous natural selection, and the large majority of those born alive are healthy in the sense that they are adapted to the environment in which they live. The primary need is for sufficient food, which requires both an increase in food supplies and limitation of numbers. Man also needs protection from certain hazards in the physical environment, particularly those which lead to exposure to infective organisms. The notable difference between human and other animal experience in relation to health results from ethical restraints which prohibit public control of reproduction. But man is uniquely educable, and can learn voluntarily to limit family size. In this way a self-imposed behavioural change may achieve the same result as the restrictions applied to other animals (p.7).

We think that McKeown is a trifle too positive here. He underestimates the hazards of the environment since his biology is too simplistic. Population regulation is not quite as neat as he suggests (see e.g. Krebs, 1972; Begon, Harper and Townsend, 1986).

    McKeown's views of natural selection and adaptation also are some-what misleading. To begin with, the *concepts* of selection and adaptation, especially the latter one, are notoriously complex. Adaptation is a *relative* notion. Organisms are adapted *to* some environmental conditions *with respect to* particular criteria, *as compared with* other organisms. Now environments are changing most of the time (on most of the time-scales one may want to consider). So organisms will need to keep track. Many biologists think that it is fairly normal for organisms to lag behind the prevailing environmental conditions (Van Valen, 1973; Dawkins and Krebs, 1979; Dawkins, 1982; Stenseth, 1985). Anyhow, there are reasons to reject McKeown's thesis that "the large majority of those born alive are healthy in the sense that they are adapted to the environment in which they live".

    McKeown envisages an *artificial*, man-made environment which ensures health for most people. Experience with husbandry allegedly suggests that the creation of such an environment is possible in principle. However, McKeown's view of husbandry is rather one-sided. Let us uncover the other side of the coin. Man has indeed tried to create artificial environments with few pathogens for many crops, and he has been successful. Pesticides were a great invention. However, nature is beginning to strike back. Pathogens tend to become resistent. In some cases, one is almost left without new *chemical* options. What should the answer be?

Presumably, one had better return to less artificial environments. That is possible only if one is willing to become less ambitious (perhaps less production, more infested specimens, more labour, etc.).

What about adaptation and disease in man? Various pathogens have almost been rendered harmless. Several countries are free of many infectious diseases which ravaged human populations in the past. But let us not become overconfident. It is conceivable that we are but experiencing a transient phenomenon. Nature may strike back in unexpected ways (cf. resistence of microorganisms to antibiotics; AIDS). There is no reason to assume that current methods to combat grand infections will remain generally successful in the long run.

Our intention is not to prophecy doom. We merely want to redress over-optimistic views. The issues involved are to a large extent undecided. By all means let us follow McKeown's lead, and try to import more ecology into medicine. But the ecology had better be good. Hypothetical environments which are populated throughout by healthy people may not be ecologically feasible.

We have one additional point about natural selection. Biologists mostly connect selection with *reproductive* survival, not survival *tout court*. One could argue that old age is not very important from the "standpoint" of selection. Whatever fate awaits you near the end will not matter, once your offspring thrives without needing you. (This is of course simplistic. We only want to make a point.) Now man has managed to extend his life-span far beyond the reproductive period. What about the diseases associated with old age? Is it reasonable to attribute them to specific environmental influences, bad habits and the like? Or should one expect that the association of old age and impaired health will remain irrespective of changes in environment and life-style which one may effectuate? There is no convincing argument to the effect that old age without health problems can be realized in principle (see Rose, 1985, for comments on relations between selection and senescence).

McKeown does acknowledge the special position of diseases associated with old age. After noting that increases of life expectation have been smaller at advanced ages than at young ages, he argues as follows.

This might be interpreted to mean that the so-called degenerative diseases cannot be expected to decline because they are determined at the time of fertilization and, being in the post-reproductive age period, are removed from the effects of natural selection.

A few decades ago it would have been difficult to reject this conclusion. However, it is now clear that some of the common causes of death in middle and late life are largely determined by the environment; for example, chronic bronchitis and coronary artery disease. But perhaps the most impressive grounds for believing that many deaths in this age period are preventable is the evidence that most cancers are due to influences which in principle could be modified (p. 24).

McKeown then comes again with an optimistic view. Any disease needs a certain genotype, but "its ill-effects are manifested only in a suitable environment" (p. 24). So removal of adverse influences involved in disorders of old age remains conceivable.

Conceivable indeed. But is it plausible? The increased life-span of man in our culture necessarily involves a prolonged exposure to prevailing environmental influences. And apparently innocuous influences may well become a threat if exposure is protracted.

So far, we have concentrated on McKeown's thesis that medicine needs to pay more attention to effects of the environment on health and disease. But personal behaviour, according to him, is at least as important as a determinant. So some additional comments are in order.

McKeown makes a distinction between (preventable) diseases associated with poverty and with affluence, respectively. The first category involves effects of environmental factors such as malnutrition, defective hygiene and bad living and working conditions. The second category is different. Personal behaviour supposedly is a major source of illness in affluent societies (cf. smoking, drinking, lack of physical activity etc.).

By and large, this is a sensible view, but we have a few critical comments. First of all, the distinction of environmental influences and influences of *personal* behaviour is a bit awkward. It may make one overlook the impact of the psychosocial *environment*. In the past few decades, this dimension of the environment has received much attention in psychosomatic medicine. We will discuss it in chapter VI. Here we only want to point out that McKeown's analysis is somehow incomplete. Perhaps he does not want to grapple with the mind-body problem and related issues, which call for attention when one takes the approach of psychosomatic medicine seriously. When he mentions the mind-body problem on p. 4 he argues briefly that "the subject need not detain us".

What about McKeown's thesis that personal behaviour is a major determinant of disease in affluent societies? We think that he may underestimate the role of environmental hazards. Consider the following line of reasoning. On p. 90 he presents data concerning smoking in relation to life expectancy. He then states:

This result can be interpreted to mean that in the past century the improvement in expectation of life of mature males *from all causes* has been reduced by at least half by smoking alone. The fact that so large a reduction has been due to a single practice suggests that in advanced countries behavioural influences are now more important than others; and since the changes in behaviour are characteristic of an affluent society, it seems permissible to conclude that diseases associated with affluence are now predominant.

Naturally, smoking is an important cause of disease. But McKeown's figures may be misleading. Life expectancy is but one criterion. What array of factors is responsible for the great variety of diseases that now characterize old age (and which often involve much suffering though they do not always greatly affect life expectancy)? It is surely possible that the physical environment plays a large though covert role. Our environment now contains great numbers of chemicals (pesticides, food additives, pollutants emitted by industry, etc.) which may affect health. It is hardly possible in practice to evaluate effects they may have, particularly long-term effects. The same goes for many synthetic drugs which are used in large quantities. Here one is faced with much ignorance.

There now are good reasons for restraint with respect to the prediction of future disease patterns. Perhaps the physical environment will increasingly become a major source of illness. For example, many potentially dangerous chemicals and viral mutants are a recent addition to the environment.

The elements of bias we uncovered in McKeown's general arguments all play a role in his view of mental illness. So let us conclude this section with some comments on this subject.

On p. 172, McKeown argues that "mental illness presents one of the most serious challenges to medical research". He then continues as follows.

But in what direction should we be looking? During man's evolution, natural selection must have restricted the frequency of genetically determined mental illness to a low level. The common diseases such as mental subnormality and the psychoses are therefore not established irreversibly at fertilization, but are due to influences acting on variable genetic material. Operationally the important consideration is not the nature of the genetic component or the balance sheet of nature and nurture; it is the feasibility of identifying and controlling the environmental influences.

In the case of mental subnormality these influences are usually, although not invariably, prenatal; in the case of the psychoses and psychoneuroses, it seems reasonable to believe as a working hypothesis that they are mainly post-natal. And just as we think of agents entering the nose and the mouth as likely to be important in respiratory and digestive diseases, so we should look to behavioural influences as the probable source of most disorders of behaviour.

There are two theses here. Firstly, McKeown thinks that genetic factors cannot be very important as determinants of mental illness. This conclusion is based on a general view of adaptation and selection which we have already criticized. Adaptation needs not be as pervasive as McKeown thinks it is! Secondly, McKeown subsequently jumps to the conclusion that "behavioural influences" are the culprit. Why could not the physical environment play an important role? Our present ignorance simply does not warrant either conclusion.

All this sounds rather negative. However, we are positive about McKeown's general ideas. He is quite clear about the limitations of his book. He only wants to indicate new directions for medicine (very sensible ones, we think), nothing more. The implementation of his ideas, *if* they are adopted, may take decades. Our comments are not meant as mere criticism, but rather as a supplement.

4.4. Facts and values: marriage or divorce?

Empirical matters were at centre stage in the previous section. Thus we have only covered one aspect of McKeown's views. How does he deal with values? We think that his general strategy is as follows. He formulates goals and values for medicine which will hardly provoke dissent. Then he asks whether we can reach the goals and, if we can, what is the most sensible way to reach them. For example, who would like to object to the following theses? Medicine should promote health for as many people as possible. Where cures are impossible, medicine will have to provide care. Prevention of diseases is at least as important as curing them once they prevail. And so on.

How can one reach such goals? McKeown's answer to *this* question is his most important contribution. He shows that medicine and health care are based on erroneous assumptions concerning empirical matters. So one is back at the facts again. But notice that the way one looks at the facts will influence one's values. Naturally, empirical matters will not make one conclude that medicine must not promote health after all. Some values are generally accepted as basic. But the adoption of more specific values, and priorities among them, will depend on how one looks at facts. *If* personal behaviour is an all-important determinant of health in our society (an assumption concerning facts which McKeown defends) one is faced with pressing questions which do involve values. How far can the medical profession go in attempts to change the life-styles of patients? The way one views *responsibilities* (of physicians and of patients) can hardly remain unaffected by ideas on the causes of disease.

Likewise, priorities will be determined, at least partly, by the estimated effectiveness of alternative strategies to deal with disease (a factual issue). If attempts to realize cures are hardly effective in some cases, both care and prevention will need to get more attention.

Notice that we are not saying that normative statements about goals and values can simply be deduced from factual statements. That would amount to a naturalistic fallacy. The situation is more complex than that. Consider the following argument scheme. Premisses: (1) One must try to reach goal G. (2) Method A leads to G in a more effective way than method B. (3) One must try to reach one's goals with relatively effective methods. Conclusion: In attempts to realize goal G, one must choose method A rather than method B. This is a valid scheme with a normative conclusion, and normative *besides* factual premisses. The interesting point is that *empirical* evidence may lead to the rejection of an empirical premiss (2), and *thereby* to the rejection of a *normative* conclusion. So there is a close connection between the domains of facts and values. But it remains possible to retain a clear *distinction* between the two domains. In this sense facts can be divorced from values so that they are "value-free".

Factual matters will affect the way one views goals and values. In some respects, the reverse is true as well. Values will affect one's approach of the domain of facts. For example, the choice of subjects for research can hardly be value-neutral. Does this show that medical science cannot be value-free? Let us be careful again. "Value-free" is an ambiguous concept. Science, medical or non-medical, is surely not value-free in the sense that it is not in any way concerned with values. Proponents of the thesis that science is value-free usually mean something else, viz. that the statements of science do not (or should not) *express* values. That is, the *truth* or *falsity* of such statements supposedly depends on nothing more than facts. Discussions about values in medicine are often confused since various meanings of "value-free" are not distinguished.

There is yet another argument designed to show that medicine cannot have an exclusively empirical domain. Central *concepts* of medicine such as "disease" *involve* values. So even the theories of medical science cannot be divorced from values.

This argument, too, is not as straightforward as a casual inspection suggests. There are many kinds of involvement. If "disease" connotes values, it does not follow that statements of medical science about disease need to *express* values. One can speak about values and still stay in the empirical domain. *Values can be mentioned in a descriptive way* (cf. comments on Pellegrino and Thomasma's views in section 2.3). The role of values in the concepts of medicine is explored in more detail in chapter IV.

All these comments hardly concern what McKeown has to say. He does not address philosophical aspects of relations between facts and values. But he seems to *presuppose* that facts and values can be kept distinct. Our comments were meant to uncover this assumption, and to show that it is reasonable if it is understood in the proper way.

5. CONCLUSIONS

All organisms represent a delicate balance of nature interacting with nurture, and genes interacting with the environment. Our species has followed a unique course due to its increasing power to manipulate the balance for better or for worse. Medicine is among the tools of manipulation. An evaluation of its roles reveals negative besides positive functions. It should be admittted that noxious effects and even malign intents, besides benefits, are features of medical practice, past and present. World-wide, medicine still serves the rich more than the poor. So it adds to social tensions and instability. It also lends itself to condoning the suppression of political dissent, in "eastern" and in "western" countries (Lüth, 1976, pp. 29-42), and it may even assist in torture (Silver, 1986). At the same time, medicine has its share in the struggle against suffering and poverty in developing countries and inequalities in health within industrialized countries, and in efforts to reduce environmental hazards. Its work for the relief of individual suffering needs no emphasis; this is what the philosophy of medicine is usually about!

Today, western civilization tends to dominate the world, and medicine mostly takes the form of western medicine with its emphasis on science and technology. This adds a new dimension to the balance of positive and negative forces. Medical science is a basis for new treatments. And technology enables us to give them a sophisticated form. But is all this really helpful? What about side-effects? What are the costs? Is there really an increased effectiveness of medical treatments? Such questions are not always easily answered.

It is easy to take sides in some of the issues we just mentioned. Medicine should not assist in torture. It should serve the poor as well as the rich. And so on. Such statements are now part and parcel of international consensus (cf. declarations of the UN, the WHO, and other international organizations). We do not want to mount a defense of the elementary values involved. We simply accept them.

Accepting values may be easy in some cases, making them operational is a different matter. It clearly goes beyond the powers of medicine and the philosophy of medicine.

We think that philosophy of medicine has a clear role vis-a-vis less clear-cut issues. For example, the appraisal of modern science and technology in medicine involves many problems with philosophical aspects. In how far are the treatments we administer effective? Are the scientific theories they presuppose consistent and well-confirmed? If positive effects on health are postulated, how is "health" defined? Such questions call for a contribution from the philosophy of science. At a more general level, philosophers of medicine should be interested in the nature of multiform interactions between medicine and culture that make medicine a positive or a negative force.

The three philosophies discussed in this chapter are all concerned with relations between medicine and culture. We have already indicated that we have a clear preference. McKeown's view seems to be the best option since it is by far the most realistic one. A more explicit comparison of the three views is given below.

We have argued that both Pellegrino and Thomasma's philosophy and anthropological medicine are culturally biased. McKeown's ideas, by contrast, may be called transculturally valid. Some explanation of terms is needed here. Our point is not that a philosophy of medicine is adequate only if it designs a medicine that can function in the same (positive) way in all cultures. Local conditions may require quite specific forms of medicine. McKeown's distinction of diseases associated with affluence and diseases associated with poverty is a case in point. Medical *practice*, at least, will have to adapt itself to cultural setting.

Does the same hold for medical *theory*? That is a moot point. McKeown's approach moves towards "transcultural validity" precisely because it tries to overcome one-sidedness associated with our culture. The Cartesian heritage has made medicine concentrate on the body, on anatomy and physiology. It has often disregarded the biological and the cultural environment. So one needs to add an ecological approach to medicine. This should give us better theories (besides better practice). Of course, one wants theories to be universal in the sense that they are valid everywhere (applying them in practice, as we saw, is a different matter). We tend to think that (western) science offers the best opportunities to develop such theories. But it is conceivable that *some* problems are better solved by cognitive styles which are foreign to us.

No integrated body of medical theory that adequately deals with effects of the environment is now available. And surely current western medicine cannot make interactions between the mental and the physical intelligible. However, alternative medicines, in other cultures but also in our own, have since long concentrated on similar issues. In the next chapter, we will analyse some forms of alternative medicine and their possible contributions to medical theory.

CHAPTER III. REGULAR *VERSUS* ALTERNATIVE MEDICINE

1. INTRODUCTION

Medicine does not take the same form in different societies. It is visibly influenced by cultural setting. Even within any particular society medicine is not a homogeneous entity, as few societies today are culturally homogeneous. There is indeed no consensus on the proper identity of medicine.

Everywhere, questions of demarcation are being debated in terms of various dichotomies: scientific medicine *versus* quackery, regular *versus* alternative medicine, reductionistic *versus* holistic medicine, etc. Such dichotomies need not coincide, so opposing parties often use different criteria in defending cherished forms of medicine (Thung, 1980). Criteria, however, are seldom made explicit beyond reference to generalities such as "adequacy". In this chapter, we will analyse specific criteria of adequacy in greater detail. First, however, we will explain the distinction of regular *versus* alternative medicine, our main subject.

The concept of regular or "orthodox" medicine commonly refers to medicine as *usually* taught in *most* medical schools (notice the qualifications). Other forms of medicine are lumped together as alternative. We will accept this distinction for the purpose of discussion; a better one does not seem to be available.

One should realize that the demarcation of regular and alternative medicine is problematic, if only because pluriformity of academic curricula is on the increase. There is no well-defined body of universally accepted medical knowledge, and there probably never was. Yet, during the late 19th and early 20th century, when applications of natural science made medicine increasingly successful, criteria for universal acceptance seemed to be forthcoming. From the end of this period stems the following authoritative dictum (Van Rijnberk, 1942, p. 14; translation ours). "Strictly speaking, one may consider the physician's training to be complete and perfect once he has learned all there is to be learned about the body. Everything besides and beyond that, is left to the physician's personal views and taste." (Van Rijnberk was a professor of physiology at Amsterdam, the Netherlands.)

Van Rijnberk's dictum soon became obsolate. The behavioural sciences entered the scene of medicine (cf. medical psychology and sociology). So did philosophy (cf. medical ethics). This helped to make the content of medical curricula increasingly variable. Medical faculties today are even offering courses on traditional medicine (India, China) or alternative medicine (Europe). There is still a common denominator which all would identify as regular or "orthodox" medicine, but it is ill-defined.

It has been thought that regular medicine may be defined as well in terms of legal recognition. We would not agree. Legal criteria *presuppose* more fundamental criteria for distinguishing regular and alternative medicine, and the law need not reflect such criteria very faithfully. Let us illustrate this by comments on the situation in our own country.

In the Netherlands, the legal definition of medical practice since 1865 covers all diagnostic and therapeutic interventions by physicians trained and licensed at universities. At the time, this was thought to guarantee homogeneous standards of theory and practice, so additional criteria were not formulated. Specifications of the medical curriculum imposed in 1916, were indeed loosened in 1968 because they were felt to be unduly restrictive. Today, course contents are allowed to vary considerably. Moreover, physicians may adhere to whatever medical theory appeals to them once they are duly licensed

In our country, the increased social acceptance of alternative medicine has recently helped to build up political pressures to change the law. The practice of healing arts outside academic medicine is soon to be legalized. In many cases social and private insurance agencies already reimburse "alternative" treatments, and an increasing number of physicians incorporate them into their practice.

In spite of such changes, physicians representing "regular" medicine remain privileged, albeit only in special risk areas (e.g. surgery, obstetrics). The distinction of regular and alternative medicine does persist, here and elsewhere. Why should that be so? In pondering over causes, one will have to concentrate on science. The crucial role of science in our civilization seems to explain the association of "regular" medicine with universities and medical schools.

Specifically, there appears to be a covert ideal of adequacy with the following features. The theories of regular medicine should be based on good science, especially natural science. They are to function as a basis for medical practice. Diseases call for scientific explanation, and once we have explanations we also have the best basis we can have for developing effective therapies. Alternative medicines represent a different ideal, or partly so; a more specific description is impossible since alternative medicines are a mixed lot.

Regular medicine's ideal of adequacy is actually treacherous. What constitutes good science? General acceptance (among scientists) is obviously a poor criterion. Scientists accept or reject theories *because* they are good or bad, so one will want to know what *their* criteria are. Unfortunately, there is much disagreement among scientists about criteria. Anyway, we need a starting point, otherwise we will be running around in circles. Therefore, "good science" in our description of regular medicine's ideal will be taken to mean "generally accepted science".

Regular (orthodox) medicine will at best qualify as a moderately adequate enterprise *if* its ideal is accepted as a standard. It contains much "theory" which does not really feature as (modern) science. Textbook material often stems from the first half of this century, or even earlier periods. The knowledge of signs and symptoms, and their association with syndromes and diseases, still is largely a store of experience passed from generation to generation. The same applies to seizable parts of our pharmacopeias, although the role of the older tinctures, tonics and pastes has dwindled rapidly during the last decades. Moreover, the theories of medicine often fail to enlighten practice. Many transactions between physician and patient are hardly affected by guidelines of science. Thus diagnoses (e.g. influenza, gastritis, migraine) and prescriptions of medicines (antibiotics, corticosteroids, benzodiazepines) often merely rest on an educated guess. For the rest, there is no warrant for much optimism about the power of science to solve major health problems in industrial countries (cf. various forms of cancer; chronic cardiovascular, neurological and locomotor diseases).

The ideal of regular medicine is scrutinized from various angles throughout this book, and regular *versus* alternative medicine therefore is a recurrent theme. In the present chapter we specifically concentrate on methodological criteria for evaluating the role of science in orthodox and alternative medicine. Anglo-Saxon philosophy of science will be an important source of inspiration. It has paid much attention to the demarcation of science and non-science, and it has since long concentrated on the development of methodological criteria for evaluating science.

We will first give some examples, in section 2, which reveal short-comings in extant general descriptions (within medicine) of science and philosophy of science. Section 3 surveys methodology from the perspective of recent Anglo-Saxon philosophy. The rest of the chapter is largely devoted to one case-study. Homeopathy was chosen ás an example for the following reasons. We wanted to begin with an easy problem. Homeopathy has its roots in an early 19th century effort to reform schools of medicine of the time. It thus shares its ancestry with present-day regular medicine, and this could facilitate comparisons. As it turned out, any adequate comparison will have to be very complex. The prospects for evaluating more outlandish forms of alternative medicine vis-a-vis normal medicine are therefore rather bleak. A brief discussion of psychic healing will underline this conclusion.

2. HOW NOT TO THINK ABOUT SCIENCE AND PHILOSOPHY

2.1. Getting science straight, the case of clinical ecology

The very suggestion that much so-called neurotic symptomatology such as *anxiety* and/or *depression* can be an ecological problem seems ridiculous. Richard Mackarness, in *Not all in the Mind,* describes an enormous catalogue of patients with seemingly severe personality disorders and depressive illnesses who have all responded to some food exclusion. However, the enquiring and critical clinician finds it very difficult to accept these observations; it just doesn't fit into the framework of conventional medicine. How can such a simple food like wheat precipitate such mental anguish as to cause an attempted suicide? ...

It is difficult to assess the claims made by Dr Mackarness in a critical and objective manner, particularly as states of mental disorientation and disturbance are almost impossible to measure objectively. ...

We simply do not know what percentage of psychological or psychiatric disturbance is precipitated by food or chemical sensitivity, but it is obviously not the complete explanation for all such problems.

The quotation is from Lewith and Kenyon's book *Clinical ecology, the treatment of ill-health caused by environmental factors* (1985, pp. 40-41), which was written for a general public. The authors work at a UK centre for the study of alternative therapies. In their book, they present clinical ecology as an alternative to conventional medicine (each with its own strengths and weaknesses). They describe themselves as the first doctors to teach the subject in the UK. The subtitle of the book adequately describes the field of clinical ecology if "environment" is taken to mean "physico-chemical and biological environment". More specifically, Lewith and Kenyon concentrate on food allergies, and on effects of pollutants in the environment. They suggest that many diseases (e.g. infantile colic; ear, nose and throat problems; eczema) respond to treatments based on an ecological approach (e.g. elimination diets), and that the orthodox medical community is not sufficiently aware of this. As the quotation shows, they are not very sure about mental illness for lack of adequate data.

Lewith and Kenyon's general point about ecology is well-taken. In conventional medicine, general practitioners and specialists alike are often not acquainted with ecology as a branch of biology because medicine, in assimilating biology, has mostly concentrated on anatomy and physiology. (Of course, there are exceptions such as epidemiology.) Ecology is indeed a relatively young branch of biology. On the contrary, ecological approaches are quite common in various forms of alternative medicine.

Before giving further comments, we want to introduce some additional material concerning food allergy. The subject plays a role in another book which appeared in the same year (1985), *Food and the gut*, edited by Hunter and Jones, with contributions by representatives of conventional medicine in the UK. It reviews recent advanced research, with much emphasis on physiology, biochemistry and immunology. But the environment has its share; after all the subject is *food* and the gut. Food intolerance (allergy; "pseudo-allergy") is among the topics considered. Recent research shows that the import of intolerance is underrated. True, cow's milk intolerance and a few other "allergies" are well-known. But other pathological conditions which sometimes involve food intolerance are usually considered to be psychosomatic disorders. Irritable bowel syndrome is an example. In many cases it does respond to treatments with elimination diets (for reference, see also Wüthrich and Hofer, 1986; Brostoff and Challacombe, 1987).

The appearance of *Food and the gut* may indicate that the tides are now turning in orthodox medicine. As yet, however, the part played by ecology remains small even when environmental factors are implicated beyond doubt. Medical researchers will often tend to concentrate on sophisticated branches of physiology. That's where the glamour is. (We do not wish to imply that physiology is unimportant; ecology and physiology are equally relevant for the understanding of allergies.)

Orthodox medicine and clinical ecology differ primarily in the weight given to the environment, to ecology. *That* difference hardly justifies the label "alternative" for clinical ecology.

The ecology theme merits emphasis because it illustrates various important points. There is much science nowadays. So much indeed that fragmentation has become unavoidable. The fact that medicine at times neglects ecology is an example of gaps in science that attend the proliferation of disciplines. Is there a remedy? Sometimes there is. Medicine draws on biology. And one can always check, from time to time, whether what passes as biology is adequate in the sense of representing real biology. One does not need sophisticated methodological criteria for that purpose. In many cases, the study of elementary biology texts will suffice.

The tendency to disregard ecology in medical research may have far-reaching consequences. For example, it facilitates the interpretation of "environment" as "psychosocial environment" (see chapter VI). The study of the environment is then implicitly relegated to psychology and social science. No wonder then that mental illness (paradigm cases, that is), in the orthodox view, gets a biological interpretation which skimps ecology. Analogous examples concerning the role of ecology in medicine are given in chapter II (section 4.3) and chapter IV (section 2.2).

So far, we have only looked at the role of science (biology) in medicine. Philosophy of science also plays its role, but it mostly does this rather surreptitiously. Some epidemiologists have recently focussed on philosophy in an explicit way. Let us see whether what they have to say will help to illuminate our main theme, orthodox *versus* alternative medicine.

2.2. *Getting philosophy straight, the case of epidemiology*

Epidemiologists are exceptionally concerned about their method of approach. In few other medical sciences is so much attention devoted to the philosophical, as opposed to the purely technical, aspect of method. The reason for this is that in epidemiology the experiment plays a relatively minor role. It is usually necessary to create a quasi-experiment out of naturally occuring phenomena. The dangers inherent in a quasi-experiment have made epidemiologists especially attentive to the logical foundations of their work. But their concern with logical rigour is narrow in a Popperian sense and could be raised to a more creative level if Popper's views were better known. Had I encountered Popper's writings earlier, I would have done many things differently (Buck, 1975, p. 160).

That is how Carol Buck started a discussion on the relevance of Popper's philosophy in epidemiology in the *International journal of epidemiology*. Before considering her approach, we will briefly put it in a broader context.

Scientists in their daily work continually frame hypotheses, they test them, and accept or reject them, give explanations, and so forth. In doing this, they mostly do not argue about the nature of hypotheses (theories, explanations) *in a general way*. They just consider some particular hypothesis and start working on it. Ask a scientist for the *precise* meaning of terms like "hypothesis", and you are likely to draw a blank. Now philosophers of science have made it their job to characterize science at this more abstract level. And not only that. They also try to develop criteria that hypotheses (theories, explanations) have to satisfy in order to be adequate. This facilitates the distinction of science and other domains, i.e. it generates *demarcation criteria*, and thereby gives the working scientist methodological tools. Or so it should. Scientists, unfortunately, seldom consult the writings of philosophers when they have to decide about any course of action, if they consult them at all. But perhaps they are right. The science philosophers write about is not always like real science.

Among the demarcation criteria, *testability* has a prominent place. Hypotheses (and theories) must be testable, otherwise they should not count as science. In early decades of this century, testability was mostly interpreted as verifiability: hypotheses are adequate only if one can *show*, in principle, that they are true *if* they are true. *This* kind of testability proved to be untenable. Interesting hypotheses often have the nasty property of being universal, i.e. they cover infinitely many instances. So they cannot be verified. Exit verifiability principle. Popper elaborated a more promising variant of the testability principle in his famous book *Logik der Forschung* (1934): falsifiability. Hypotheses are adequate only if it is possible to *show* that they are false *if* they are false. In scientific research one should aim at the refutation of hypotheses since one can only approach the truth by eliminating what is false. In the process of refutation one uses deductive logic, which goes with certainty from accepted premises to conclusions. (Popper does not accept any kind of inductive logic.) How does one arrive at new hypotheses? Popper's answer is, through bold conjectures. Logic and methodology will not be helpful here. The deductive approach, applied to hypotheses once they are elaborated, is to ensure *objectivity* and *rationality* in science.

It is this philosophy which Buck wants to introduce in epidemiology. She argues her case with many examples. Let us consider some of them. On p. 162 she states that one can have two different reasons for replication in epidemiological studies:

The first has to do with the need to confirm observations because the play of chance can create spurious associations. ... Here, then, is one justification for replication of an epidemiological study. When this is the purpose, the replication should be as exact a reproduction as is possible of the original study. ...

The second reason for replication is deductively much more powerful in Popper's sense. Suppose one hypothesizes that an endogenous degenerative process plays a major causative role in a disease. A positive association of the disease with age would support the hypothesis. But the identification of a population in which age had no relationship to the disease would suggest the alternative hypothesis that it is caused by prolonged exposure to an exogenous factor, present in some populations and absent in others. The greater one's age, the longer would be one's exposure to the factor if it were present. Repetition of a study that showed age-dependency thus should be made in a population as different as possible from the original one. Replication then becomes an attempt at refutation.

Would it indeed be possible to falsify a hypothesis in this way? We doubt it. Our first reaction to the hypotheses would be that we want clarification. Are terms like "endogenous" and "exogenous" really clear? They might well cover a lot of theory that needs a separate evaluation. More importantly, would the difference between populations which Buck mentions allow the rejection of any hypothesis? If "prolonged exposure to an exogenous factor" is implicated, one should not be surprised if there is a relationship between disease incidence and age. But even apparently different populations may well be similar *with respect to* an important "endogenous" factor. Thus a "positive" outcome (no difference between populations) will not tell us much. Likewise, differences between populations could reflect differences in "endogenous" factors. Refutation (falsification) cannot be as straightforward as Buck suggests.

Buck's example is really rather complex, more complex than she suggests. And the kind of complexity involved is not typical of the example, it is typical of science. For this reason, there are *always* many options when evidence contradicts an hypothesis. One can blame the hypothesis. Or one can blame the evidence (confounding factor, failure of equipment, etc.), or background theory (there always is a lot of theory at the background). Lastly, the decision can be postponed. Philosophers know about this. Most of them have by now rejected Popper's approach. Buck, even though she may not be acquainted with recent philosophy, should know better. She is familiar with science.

All the examples which Buck gives will easily lead to the same conclusion. However, we want to quote one more example to make a different point.

Epidemiologists are sometimes asked why John Snow invariably receives the greatest prominence in any historical account of their subject. The simple reason is that he was successful in establishing a causal mechanism for an important disease and that he accomplished this by methods that we now call epidemiological. What needs more emphasis in our praise of Snow is the consistently deductive nature of his approach. When he gathered new data it was always for the purpose of testing a highly specific prediction of his hypothesis. The incident of the Broad Street pump in Soho has become famous as one of the first experiments in epidemiology. ... Snow's recommendations that the pump be closed may not have been viewed by him as an experiment so much as a mandatory action. Having shown that people living beyond the area of the pump who had used its water were attacked, while residents of the area who did not drink from it were spared, what other action could have been recommended? (p. 163).

We agree that this is a rather nice example which shows that research will often benefit from rigorous hypothesis testing involving deduction. So what Popper has said is not *merely* nonsense. At the same time, the example has strongly un-Popperian features. Snow seems to have arrived at his hypothesis by a clever piece of classical inductive reasoning! Can you find any systematic difference between people who get a disease and others? If there is a difference between them in some particular factor, assume that this factor has a causative role. Epidemiologists are all the time proceeding in this manner. Popper should not be pleased with the example.

Buck's article was followed by brief rebuttals (Davies, 1975; Smith, 1975; Jacobsen, 1976). Later on, Maclure (1985) resumed the defense of Popperian epidemiology (for comments, see Weed and Trock, 1986; Maclure, 1986). Philosophers would not be pleased with the whole discussion. Fortunately, other epidemiologists, Susser (1986) and Marmot (1976, 1986), have presented a much better account of philosophical matters in epidemiology (see also Maclure, 1987, and Susser, 1987, for comments). Marmot's basic sympathies are with the philosophers Kuhn and Lakatos. Kuhn (1962) is among those who have cast doubts on the rationality of science. (Remember that Popper's falsifiability criterion was meant to function as a criterion of rationality.) Lakatos' (1972) views are somewhere between those of Popper and Kuhn. Marmot argues that epidemiology is indeed not characterized by the rigorous kind of rationality that Popper advocates. One often clings to cherished hypotheses in the face of adverse evidence. And that is a good thing. Hypotheses and theories without protection are apt to be rejected for the wrong reasons.

Alternative medicine is not an issue in any of the articles we mentioned. For us, the connection is as follows. Methodological criteria such as testability are meant to be a hall-mark of rationality in science. Thus falsifiability, for Popper, represented the water-shed between science and metaphysics. If our preliminary definition of orthodox *versus* alternative medicine in terms of science is sensible, the methodology used to characterize science will make a difference. In the next section we will survey characterizations given by Anglo-Saxon philosophy.

3. WHAT IS SPECIAL ABOUT SCIENCE

There are many kinds of knowledge. In daily life, one knows about the wheather and about where to get groceries. One is pretty sure about *that* kind of thing, but as one turns to more general matters the situation is different. Take religion. There are those who seem to know about the meaning of life, about God. But there is no *common* knowledge of such things. Science is apparently different. It represents a special kind of knowledge which people can share. Its method is objective, it is designed to uncover the truth. Science is *rational*.

This picture of science will not do anymore. The examples in the previous section indicated, in a preliminary way, what is wrong with it. Science, in the quest for knowledge, proceeds by testing hypotheses. So hypotheses must be testable to begin with. But there are problems with the testability principle.

Other methodological principles characterizing the method of science are not straightforward either. They are briefly surveyed in this section. We begin with additional comments on testability.

Philosophers do agree nowadays that rigorous falsification is as unattainable as verification (see e.g. Harding, 1976). Testability at best amounts to the weak requirement that theories be accessible to (dis)confirmation through evidence. Now confirmation or disconfirmation of some theory is not very helpful unless it discriminates between the theory and (some of) its rivals. This spells additional problems for testability as a criterion of rationality. Discrimination is seldom simple. Take smoking and lung cancer. Since decades, there is epidemiological evidence, experimental evidence based on work with mice, and so forth. The evidence could hardly be more impressive. But in individual cases it remains possible to question the causal relationship. Bronchial carcinoma did occur, though as a rare disease, before cigarettes were invented and today even heavy smokers may remain unaffected. By now there is broad agreement that smoking is bad, but the battle was not won overnight and it involved much weighing of evidence.

Kuhn, in his classic *The structure of scientific revolutions (1962)*, argued that major changes in science cannot be explained at all in terms of rational procedures. "Paradigms" are tough, and so they should be if there is to be order in science. They will not fade away as evidence accumulates against them. Sheer evidence will not make defenders of a new paradigm win. They need power rather than arguments.

Kuhn's paradigms include basic presuppositons of science, theories, methods, and other items shielded from critical tests. Paradigms define what is to count as admissible evidence. And they colour the meanings of concepts we use for describing evidence. So there cannot be a standard for comparing theories in different paradigms. Any term shared by such theories should actually stand for different concepts. In other words, *meaning variance* will hinder a rational assessment of the merits of rival theories since theories are *incommensurable.*

Kuhn's stance did not remain unchallenged. An array of models for scientific change, both "rational" and "non-rational" ones, was developed by others (see Newton-Smith, 1981). Rationality proved to be a rather untractable theme (for recent comments om rationality and relativism see Hollis and Lukes, 1982; Siegel, 1985; Leplin, 1986; Wachbroit, 1987; Winch, 1987).

Other themes now share rationality's fate. For example, realism is a hotly contested issue (see e.g. Van Fraassen, 1980; Cartwright, 1983; Hacking, 1983; Churchland and Hooker, 1985; Rescher, 1987). Will scientific theories tell us what the world is like? It would be unwise to regard any theory as an ultimate source of truth. But the more modest claim that successive theories bring us closer to the truth is now disputed as well. It would imply that science is progressive in the sense that it gradually approximates the truth, and "truth" is a problematic notion.

Realism, in one sense, implies that adequate theories say what the world is like (in some respects). Truth may thus be regarded as correspondence between propositions and states of affairs. However, it is not easy to specify what "correspondence" means. Many philosophers have therefore rejected the *correspondence theory of truth* in favour of a *coherence theory*. They argue that coherence among our beliefs is the ultimate criterion for regarding them as true.

Putnam (1981) is among the defenders of the coherence theory. He argues that notions like "truth" only make sense *within* the framework of coherent beliefs we happen to adopt. He calls himself a realist, though one of a particular sort. His "internal realism" ("internalism") is invested with a modesty that comes close to relativism. As Hacking (1983, p. 128) puts it, "We are left with no external way to evaluate our own tradition, but why should we want that?" Putnam and Hacking, for that matter, would not call themselves relativists.

The picture of philosophy given so far will suffice as a warning against any rash evaluation of theories, e.g., on the basis of testability. The testability criterion is but one tool of methodology. Will other criteria be more helpful? Clarity of concepts is an obvious candidate. Obscure theories must be relegated to the realm of bad science, or non-science. It is not easy, however, to articulate a sensible criterion of clarity because there is no generally accepted theory of meaning.

What about explanatory power as a criterion of rationality? Theories that cannot be used for explanations will not be regarded as scientific. But this criterion again is not as clear-cut as it seems. Let us consider the fate of the concept of explanation in the course of modern Anglo-Saxon philosophy. The year of 1948 is a suitable starting point. Hempel and Oppenheim then published their famous account of scientific explanation. They regarded explanations as arguments, in which a statement describing the event to be explained appears as the conclusion. Two kinds of premisses are needed: laws, and statements describing initial conditions. Explanatory arguments, in order to be acceptable, must satisfy criteria for sound argumentation. Thus premisses must be true or well-confirmed and the conclusion must logically follow from the premisses.

Scientific explanations seldom satisfy such criteria in a rigorous way. Consider the connection between syphilis and dementia paralytica (DP), which many philosophers have discussed (e.g. Stegmüller, 1969). People contracting syphilis will sometimes develop DP later on if the disease is not successfully treated. It is sensible to explain DP in terms of prior syphilis. Under Hempel and Oppenheim's model this explanation has the following structure. Premisses: (1) If x contracts syphilis at $t(i)$ and certain conditions C are satisfied, x developes DP at $t(i+j)$. (2) John contracts syphilis at $t(1)$. (3) Conditions C are satisfied (in John's case). Conclusion: John develops DP at $t(1+j)$.

This is a logically valid argument. But its weakness is clear. Conditions *C* cannot be adequately specified. So the "law" (premiss 1), although it has explanatory force, is vague. Premisse (3), for the same reason, can only be inferred from the conclusion (together with independent evidence to the effect that other cases of DP were preceded by syphilis). Nonrigorous explanations of this kind are very common, not in the least in medical science.

Hempel and Oppenheim's model is now widely regarded as inadequate. Many alternatives have been proposed (see e.g. Salmon, 1984), but none of them is generally accepted. Various authors (e.g. Van Fraassen, 1980) have argued that classical approaches like that of Hempel and Oppenheim fail because they disregard the *pragmatic context*. The adequacy of explanations will depend on how one wants to use them. Many "conditions" (cf. the syphilis example) will affect any phenomenon which calls for explanation. In giving explanations, one will emphasize conditions deemed relevant in view of context-dependent goals.

4. INTERLUDE: HOW TO PROCEED?

The survey given in the previous section shows that methodology, like science, is subject to change. Methodology, too, has its controversies. Moreover, methodological tools developed by philosophers need not coincide with the tools used by scientists (and scientists need not agree about methodology either). All this will make the methodological evaluation of alternative *versus* regular medicine rather complex. But we are still optimistic. Let us explain the spirit of the strategy we intend to follow by some examples.

1. Consider again Hempel and Oppenheim's model of scientific explanation. Let it be granted that it is inadequate in the sense that it leads to the rejection of some intuitively sensible explanations, and to the acceptance of some nonsensical explanations. Should this make the model worthless? We do not think so. Scientists giving explanations will anyhow use arguments. Some of them may violate elementary logic. So one may be justified in rejecting them. This clearly agrees with the gist of Hempel and Oppenheim's non-pragmatic approach. If one rejects an explanation because its logic is bad (or because it does not even covertly refer to some general

statement) one does not need to consider pragmatics. Perhaps a generally valid explication of "explanation" (if possible at all) should cover pragmatics. This does not imply, however, that pragmatics must enter the scene whenever explanations are scrutinized. Models *in science* with obvious "shortcomings" (cf. assumptions being violated under certain conditions) are often freely used for specific purposes in a particular context. Why should one not do the same with philosophical models? Philosophers have a tendency to engage in never-ending quests for perfection, even though they are acquainted with the phenomenon of context-dependence. This hampers application of their results in the setting of scientific practice.

2. No philosopher nowadays will argue that falsification (in a strict sense of the term) of any hypothesis or theory is possible. Falsifiability, like verifiability, is simply an undefensible version of testability. Many evolutionary biologists none the less claim that application of this principle should lead to the rejection of various biological theories. Few biologists bother about recent philosophy of science, and the same goes for medical professionals (see Sassower and Grodin, 1987). Those who bother mostly concentrate on Popper's falsifiability principle, which they simplify enormously. This is clearly a situation that calls for correction of the scientist's methods (and the methodology they represent) by methodology developed in philosophy.

3. Philosophers arguing about testability often do not consider the intricacies involved in actual tests of hypotheses in science. They mostly work with simplified reconstructions of hypotheses and theories. In principle, therefore, the philosopher's methodology may be challenged by appeals to scientific practice.

The moral of the examples will be clear. If one is to make sense of demarcation problems, one must acquaint oneself with various domains of science and philosophy. Common sense *cum* elementary philosophy will then permit much useful work. The analysis of any problem, within science or philosophy, can in principle continue *ad infinitum*. In practice one will have to stop somewhere, and we will often stop rather soon. Analyses pushed beyond a certain limit tend to be self-defeating. They will easily make one forget the context in which the problem at issue arose.

Homeopathy will be our main example of alternative medicine. Our analyses will reveal that standpoints taken towards homeopathy mostly involve a mixture of science and philosophy (methodology). Our first task, then, is to disentangle issues that belong to different domains of knowledge.

5. THE SCIENTIFIC STATUS OF HOMEOPATHY

5.1. Preview

Homeopathy is among the major variants of alternative medicine. Its adherents regard its theoretical principles as well-confirmed, and its practice as benificial. Some representatives of orthodox medicine, however, brand it as pure superstition. We shall comment on its merits by an analysis of representative literature, especially the books by Vithoulkas (1980) and Mössinger (1984) in defense of homeopathy.

The basic theory of homeopathy was developed by Samuel Hahnemann in the last part of the 18th century. He formulated its central principle as follows. "Similia similibus curentur", or like should cure like, "Any substance which can produce a totality of symptoms in a healthy human being can cure that totality of symptoms in a sick human being" (Vithoulkas, p. 92).

Homeopathic drugs are mostly used in high dilutions which are prepared through a special process, "potentization" (for details, see Vithoulkas). Surprisingly, effectiveness is supposed to increase with dilution. Thus preparations which are diluted "far beyond Avogadro's number" (which do not contain any drug molecule) should be extremely potent. Conventional science, of course, is not very enthousiastic about potentization. For that matter, neither are some homeopaths.

Among homeopaths, opinions over medical theory are divided. Some of them, e.g. Vithoulkas (1980), defend a grand theory as an alternative of orthodox-medical science. Others, notably Mössinger (1984), oppose undue emphasis on theory, orthodox or alternative. Vithoulkas does not concentrate on philosophy (but he gives a detailed account of the methods used in homeopathic practice). Mössinger explicitly bases his views on philosophical criteria. He tries to bridge the gap between homeopathy and orthodox medicine by investigating effects of homeopathic drugs with conventional methodology.

We will first evaluate the general viewpoints developed by Vithoulkas and Mössinger, and then consider potentization and the *similia* principle. Lastly, we will put tests of homeopathic drugs in a methodological perspective.

5.2. Homeopathy as a grand theory: Vithoulkas

The similia principle and the rule of potentization function as "low level" empirical generalizations. Vithoulkas obviously aims at explaining such generalizations by a higher level theory. We will analyse one representative passage from his work as an example. The following quotation (pp. 87-91) shows how he explains the similia principle.

The *dynamic plane* is the plane of the essence of life, the plane on which disease originates, as well as the plane of origin of the defense mechanism. It is not as if the dynamic plane is a separate fourth level of the organism. [Vithoulkas distinguishes three levels, the mental, the emotional and the physical.] Rather, it permeates all levels, is prior to them, and interacts with them. It has exactly the same relationship to the physical body as electromagnetic fields have to matter. ... ... the vital force or dynamic plane interacts intimately with all three levels. Whenever an organism receives a stimulus through one of its three levels of reception, the effect is responded to initially by the electrodynamic field (or vital force) and is then distributed to the three levels according to the strength of the stimulus and the degree of resistance of the organism.

Modern concepts of cybernetics demonstrate a fundamental principle which applies to the human organism as well as to the other systems: *any highly organized system reacts to stress always by producing the best possible response of which it is capable in the moment.* In the human being, this means that the defense mechanism makes the best possible respons to the morbific stimulus, given the state of health in the moment and the intensity of the stress.

When disease occurs, the first disturbance occurs on the dynamic electromagnetic field of the body, which then brings into play the defense mechanism... .

Symptoms and signs are the only way we have of perceiving the workings of the defense mechanism. It is acting in the best possible manner for the benefit of the organism; for this reason, the symptoms and signs produced are actually attempts on the part of the organism to heal itself... .

To affect directly the dynamic plane, we must find a substance similar enough to the resultant frequency of the dynamic plane to produce resonance. Since the defense mechanism's only manifestation perceptible to our senses is the symptoms and signs of the person, it follows that we must seek a substance which can produce in the human organism a similar totality of symptoms and signs. If a substance is capable of producing a similar symptom picture in a healthy organism, then the likelihood of its vibration rate being very close to the resultant frequency of the diseased organism is good, and therefore a powerful strengthening of the defense mechanism can occur - through the principle of resonance.

Could Vithoulkas' theory qualify as science? We will try to answer this question by concentrating on two methodological principles, the demand that theories be somehow testable and that their concepts be clear.

If Vithoulkas' theory is to be testable, links between theory and empirical data must of course (in principle) permit answers to questions like "What *reasons* do you have for attributing certain symptom pictures to certain alterations in the dynamic electromagnetic field of the body?". Vithoulkas does not seem to give any clear answer. True, he mentions various sources of evidence (e.g. general considerations from physics, Kirlian photography), but he does not explain how such evidence will connect the hypothesized dynamic plane with the world of symptoms.

Some of his statements suggest that he himself does not regard his theory as a testable one. Consider: "Symptoms and signs are the only way we have of perceiving the workings of the defense mechanism. .... Since the defense mechanism's only manifestation perceptible to our senses is the symptoms and signs of the person, it follows that we must seek a substance which can produce ... a similar totality of symptoms and signs." These statements point to a covert methodology which we do not understand. Does our inability to *perceive* the inner workings of the defense mechanism imply that we cannot *infer* how it works either? If that is true, how can one ever know that Vithoulkas' theory makes sense? If inferences are actually allowed, why should perceptible symptoms and signs be all-important in medical interventions? There is a tension here that calls for an articulation of methodology. Vithoulkas, unfortunately, does not bother to explain his methodology. One is left, then, with questions which only Vithoulkas can answer. The burden of proof is his. Notice the we did *not* criticize him on the basis of any *particular* principle of testability. We are indeed following the approach outlined in section 4.

Vithoulkas' concepts also call for comments. "Dynamic plane (vital force)", for example, remains a rather elusive notion. At the same time it supposedly fits conventional physical theory (cf. "dynamic electromagnetic field"). Again, we are confronted with ambiguity. If notions from physics do have a role to play it would be reasonable to ask for more clarity. If physics is not really at issue, analogies of its concepts had better be left out to avoid confusion.

So far, we merely gave negative comments. Let us continue in a more positive vein. We think that Vithoulkas' theory could develop into a more plausible one if it is reconstructed as a heuristic device. Consider the following line of reasoning. "Whenever an organism x contracts a disease y, x will exhibit symptoms sy reflecting responses z, and z will promote recovery. When a drug d produces sy, and d is given to x, d will facilitate z and so promote recovery as well." The first statement is somewhat vague but it may generate well-confirmed theses of orthodox medicine after elaboration. The second thesis is the similia principle together with a higher level statement (*d* will facilitate *z*) meant to explain it. The inference of this statement is not totally unfounded. The states of organisms produced by *y* and *d* are by assumption similar with respect to symptoms. So it is reasonable to expect other similarities, e.g. at the level of responses *z*. Confirmation of the similarity in symptoms, and identification of *z* in particular cases, naturally would be helpful.

In short, there is no reason to accept Vithoulkas' higher-level theory as he formulates it. But it may be acceptable as a starting point for further research after reconstruction. (Our reconstruction is rather simplistic and only serves to indicate a feasible kind of approach.) True, even then its explanatory force is weak in view of the postulated unknown entities. But the effects of many traditional allopathic drugs also cannot be explained, if their effectivity has been established at all. Provisional theories are neither non-science nor bad science as long as we recognize them for what they are.

5.3. Against theory: Mössinger

Mössinger, unlike Vithoulkas, explicitly discusses the impact of philosopy on medicine. Medical therapy rather than theory is his main concern (Vithoulkas considers both). Differences in the general views of the two authors, therefore, partly reflect a difference in purpose.

The distinction of rational ("rationale" or "rationelle") and empirical ("empirische") approaches to therapy is at the centre of Mössinger's philosophy. He regards current scientific medicine as overly rationalistic. Homeopathy, according to him, deals with the empirical domain in a more adequate way.

Mössinger's distinction suggests that he is dealing with classic philosophical positions known as rationalism and empiricism. We are not sure if that is what he means, since he does not explain his central concepts in detail. You will have to understand them from the following summary of his philosophical views.

Man has two quite different knowledge systems, rational thought and *experience* ("Erfahrung"). Phylogenetically, rational thought is a recent acquirement. *Experience* is much older. It is essentially a trial and error system that cannot be fathomed by conscious thought. Selection ensured that an invaluable store of knowledge was accumulated by *experience*. Prior to the advent of modern natural science, such knowledge formed the core of medical therapy (cf. medicinal herbs). Modern science initially enriched medicine by putting knowledge acquisition in a more rigorous framework. Bacon, Hahnemann and Virchow, by their emphasis on the empirical domain, represent what is best in its tradition. They appreciated the data accumulated by empirical medicine. Hypothesis and experiment, of course, are essential for the progress of medicine, but they must remain in close contact with *experience*. Unfortunately medical theory, however useful potentially, was soon given too central a position. Nowadays, regular medicine tries to give therapy a rational foundation divorced from *experience*. It cannot anymore deliver the goods it promises since any theories of man (indeed, life) have fundamental limitations: life is ultimately mysterious. The valuables of traditonal medicine are now almost disregarded in circles of physicians inspired by natural science. Homeopathy aims at restoring the balance. It emphasizes individual medical therapy ("individuelle Arznei-therapie"), and it concentrates on *experience*.

According to Mössinger, rational thought and *experience* both have their strengths and weaknesses. *Experience* has its limitations, so it need to be supplemented by rational thought. At the same time a rational approach tends to lead experience astray.

Mössinger emphatically wants to integrate homeopathy and regular medicine. As a practicing homeopath, for example, he has conducted many controlled experiments designed to test effects of homeopathic drugs (see section 5.5). With respect to homeopathy's empirical generalizations he is rather critical. He regards the similia principle as a useful heuristic device, but he warns against accepting it as a dogma. He does not regard potentization as a sensible principle. It stems from a vitalistic philosophy which has not been confirmed by adequate empirical data.

Before commenting on Mössinger's views, we want to emphasize their context-dependency. He clearly states that his philosophy is not meant to bear on natural science as such. The methodology he advocates is only destined to serve medical practice.

It is not easy to evaluate Mössinger's general views since he has not developed them in great detail. (The greater part of his book is devoted to case studies, and controlled experimentation.) We like the atmosphere of his approach, but we think that his view of *experience* needs correction. What about the possibility that *experience* may submerge us into tenacious superstition, even throughout long periods of history? Naturally, one should not adopt a simplistic, negative attitude towards the era preceding the rise of modern science. Philosophy inspired by history will tell us that the Dark Ages were not simply *dark* ages. But the opposite simplism will not do either.

Mössinger justifies his positive attitude towards *experience* in terms of evolutionary epistemology, which recently gained influence especially in Europe (for references see Mössinger, 1984 and Van der Steen, 1986b). Evolutionary epistemologists roughly reason as follows. Natural selection channelled the evolution of knowledge systems so that their adaptiveness is ensured. Knowing the truth is adaptive. Cognitive faculties, therefore, can ultimately be trusted. We regard this as an extremely naive stance. Firstly, evolutionary biologists have never really studied *data* concerning cognitive processing. Evolutionary epistemology presupposes an integration of cognitive psychology and evolutionary biology where none exists. Secondly, evolutionary epistemologists almost treat adaptation as a *criterion* of truth. That will not do because it makes their central thesis (knowing the truth is adaptive) vacuous (for further details see Van der Steen, 1986b; Ruse, 1986, and Bradie, 1986, are more positive about evolutionary epistemology). We do not want to deny the import of *experience*, but we cannot accept Mössinger's defense of it.

If one wants to understand knowledge and cognitive abilities in terms of empirical science, one must turn to psychology rather than evolutionary biology. Psychologists know about our limitations. In daily life, facts are unknowingly distorted to make them fit preconceptions, motives are "rationalized" to justify behaviour, events witnessed are seldom reported with any accuracy, and so on. Science, pure or applied, is no exception. *Judgement research*, in the last few decades, has given a shocking picture of bias even in scientific work; for a review, see Faust (1984).

One of Faust's points is that man cannot handle complex information. For example, physicians will commit many errors of reasoning when they base diagnoses on a great variety of data. Faust concludes that science is using an inadequate methodology. The methodology as such may make sense, but we do not have the cognitive resources to make it work. Therefore, mechanical procedures had better take the place of human decisions in complex situations. Faust's view accords with recent developments in medicine. The emphasis has shifted from "clinical intuition" to more "objective" methods, and then to a more mechanical kind of objectivity (cf. medical decision making; see also Chapter IV).

The findings of psychology seem to suggest that this is a good development. At first sight, therefore, one would think that there is no place for Mössinger's views in modern medicine. Clinical intuition,which should resemble his *experience*, is a poor basis for medical decisions. More rational judgements also have their limitations (Mössinger is right here), but objective methods are now available.

This view, however, is dangerously one-sided. There is another side to the coin of *experience*. Take field biology. Experienced naturalists are able to recognise a species of bird where laymen hardly see more than a moving spot. They make mistakes, but their ability to come up with correct identifications is beyond doubt. *Experience*, in this case, is sometimes superior to the "rational" use of identification keys. The keys may be adequate, but one may fail to use them correctly. Clinical intuition, likewise, may have its strengths. Judgement research as reviewed by Faust does not seem to cover all the knowledge which may play a role in clinical experience. One might be tempted to assume that medical decision making (say, as implemented on computers) avoids the shortcomings both of clinical intuition and of rational methods of a more "primitive" sort. Computerized decision programs do facilitate the processing of complex data and they promote precision. They give concrete form to formal aspects of rationality. However, one will need to make choices concerning the variables to be taken into account, and weights will have to be attached to specific factors. Such choices inevitably involve personal preferences and experience. Formalized methods are useful, but one always needs something more (cf. Wartofsky, 1986).

Mainstream philosophy of science, unfortunately, has not paid much attention to the issue we have been considering. An "analytical" attitude has mostly dominated the scene. The views of various philosophers who have worked outside the mainstream (e.g. Polanyi, 1958; Polanyi and Prosch, 1975) show more resemblance with Mössinger's work. Polanyi's "tacit knowledge" strikingly resembles Mössinger's *experience*. Anyway, Mössinger's views merit serious appraisal. They indicate that common methodologies now defended, be it in philosophy or in medicine, may need substantial revisions.

5.4. Potentization and the similia principle

Potentization is doubtless homeopathy's most problematic tenet. Some homeopaths do not accept it themselves (cf. Mössinger's views). Rejection of potentization, of course, need not lead to the rejection of homeopathy as a whole.

Could preparations that do not contain any drug molecule be effective? Current chemical theory seems to warrant outright rejection of this idea. Homeopaths could of course reply that potentization activates vital forces that cannot possibly be detected by chemistry. That would amount to a kind of vitalism that runs the risk of becoming untestable.

However, homeopaths could also aim at an explanation of potentization in terms of chemistry. Suppose that empirical evidence reveals that some drugs are indeed effective at extreme dilutions (without any molecule of the drug). Some homeopaths have argued that this may be explained by changes of the solvent (e.g., water) under potentization; drugs could somehow transfer their effectivity to the solvent. The postulated transference has received some support from recent studies with methods of ordinary physics and chemistry (Sacks, 1983; see also references in Reilly et al., 1986), but further confirmation is badly needed. Sacks' work at any rate cannot be accepted in its present form. He claims that analysis with nuclear magnetic resonance has shown that spectra of normal water and water used in potentization are different. But his results can hardly be evaluated since basic data are poorly represented in the article.

Is the transfer phenomenon incompatible with chemical theory? "Incompatibility" could mean various things in this context. The following interpretations are possible. (i) Current chemical theory implies that the transfer phenomenon is impossible. (ii) Any chemical theory, present or future, implies that the transfer phenomenon is impossible. (iii) It is impossible to derive the transfer phenomenon from current chemical theory. (iv) Idem, from any theory, present or future. (v) Nobody has succeeded in deriving the transfer phenomenon from chemical theory. Etc.

Obviously, statement (v) is true. Any of the other statements is open to doubt. Statements (ii) and (iv) are plainly unreasonable. However, (v) supports (iii), and (iii) with suitably chosen additional assumptions may support (i). The latter statement warrants the rejection of specific activities of some homeopathic prescriptions, since chemical theory is well-confirmed. But in principle the claimed activities could also come to be well-confirmed. They could even lead to a revision of chemical theory. The upshot of our analysis is that the existence of potentization effects has not been disproved. Proof and disproof in a strict and definitive sense do not have a place in empirical science (see section 3). As yet, however, homeopathy's claims concerning potentization are outlandish. And chemical theory constitutes *prima facie* evidence against them. So it is reasonable to ask for extensive and solid confirmation.

The similia principle need not be affected by these considerations. It may not be inconsistent with any accepted theory. What about the evidence? Thousands of homeopathic physicians regard it as well-confirmed. Their conclusion, however, is based on their experience with homeopathic prescriptions which seldom qualifies as controlled experimentation. This, of course, will sustain critical reactions on the part of conventional medicine.

There is yet a more fundamental reason for criticism. Homeopaths, when testing the principle, often do not distinguish between its separate components. Even a well-controlled observation to the effect that substance *S* cures symptoms *Sy(1,2...n)* in sick human beings, does not validate the principle. The premiss "*S* induces symptoms *Sy(1,2...n)* in healthy human beings", must be confirmed as well. But in many cases this premiss still rests on the original descriptions of Hahnemann (1842, see Vithoulkas, 1980), or on later "provings" based on Hahnemann's methodology (see Bodman, 1968). The descriptions hardly fit present-day medical symptomatology. Homeopathic physicians describing the effects of drugs on normal persons, will adopt the language of individual "provers" (often laymen), which not surprisingly is much the same as in Hahnemann's days. True, they may use fairly orthodox standardized clinical language to report clinical cases (e.g. Griggs, 1968). But they do not pay attention to the fact that the two languagues don't match. So their "confirmation" of the premiss mentioned above will not easily be accepted by the orthodox scientific community.

As yet, we assume that the principle needs further confirmation. For the rest, its rejection would by no means imply that homeopathic drugs cannot be effective.

5.5. Testing effects of homeopathic drugs

Are homeopathic drugs effective in cases where homeopathic physicians choose to prescribe them? This question has been researched very often, allegedly with positive results, but the force of the tests is weak according to standards defended (though often violated) in orthodox medicine. Many homeopaths have argued that standard methods of testing are inapplicable in this case since homeopathy emphasizes the *uniqueness* of patients and allows no controlled experimentation.

Representatives of orthodox medicine would argue that controlled experimentation is indispensable for the assessment of medical treatments.

The point is that effects observed after some medical intervention, e.g. treatment with drugs, need not be caused by the intervention as such. Many other factors in the therapeutic setting may be effective as well. That is, apparent effects of an intervention may represent *placebo effects*. So one needs controlled experimentation to evaluate the efficacy of interventions. Effects of ordinary treatments must be compared with effects produced by treatments without the therapeutic ingredient assumed to be effective (placebo treatments). The patients evidently must not know what treatment they are getting. But that is not enough. The physicians prescribing treatments must be ignorant as well, one needs "double-blind" studies.

Various investigators have recently argued that homeopathic drugs can be investigated with normal double-blind investigation. Mössinger (1984), the homeopath, is among them. He analysed homeopathic drugs in various experiments, some of them single-blind (patients ignorant), others double-blind (both patients and physicians ignorant). (Vithoulkas also mentions the necessity of double-blind experimentation, but he does not describe research with this method.) The results suggest that one should take homeopathic drugs seriously. As yet, the number of controlled trials involving homeopathic drugs is small. Clear effects involving a rigorous design were established in a few cases only (e.g. the study of Gibson et al., 1980, on rheumatoid arthritis, and that of Reilly et al., 1986, on hayfever). Various other studies are not decisive in view of methodological pitfalls (see the review by Scofield, 1984).

Future research with double-blind experiments could show that homeopathic drugs may be effective according to standards accepted by orthodox medicine. As long as *this* kind of confirmation is scarce, cures resulting from homeopathic treatment are likely to be explained as placebo effects in the context of modern medicine. Placebo's do affect conditions which are common among the clients of homeopathy: arthritis, migraine, gastritis, asthma and other ill-defined chronic diseases and some self-limited acute conditons like tonsillitis, otitis media.

If controlled experimentation is indeed taken to be the decisive criterion for evaluating drugs, homeopathy now compares poorly with allopathic medicine. So do other varieties of alternative medicine (Maddocks, 1985; Stalker and Glymour, 1985). However, the situation is not quite that simple. Controlled experimentation faces problems of its own. The discussion in the next section will show that this calls for a revision of some of our conclusions.

5.6. The vagaries of controlled experimentation

Controlled experimentation is nowadays a widely accepted criterion for the assessment of drug effects. The law in many countries accepts it as a basis for decisions on the admission or rejection of new preparations. Recent German law, however, is more lenient. Germany's decision not to honour current international standards was motivated by methodological research on the value of controlled experimentation. Kienle, who worked at the Herdegger Institute before his death in 1982, played a dominant role in the investigations. The results of the relevant analyses are surveyed in Kienle and Burkhardt (1983). The following succinct summary gives a general idea of their central theses.

Any controlled experiment will have to focus on effects of a drug on one or a few variables, preferably measurable ones, regarded as adequate criteria for some disease. Conclusions based on such experiments naturally presuppose that the criteria are valid indicators. Unfortunately, their validity is mostly problematic for various reasons. Experiments are adequate only if the experimental group and the control group are random samples from a well-defined statistical population. In most cases, it is impossible to realize this set-up (again, for many different reasons). Moreover, the population studied will seldom coincide with the population one is actually interested in. General statements inferred from experiments, therefore, will almost always involve many additional assumptions. Heterogeneity among patients is another cause for worry. Even under correct randomization, covert interactions among factors that affect the disease studied may lead to erroneous conclusions. And so on, and so forth.

The authors thoroughly analysed a great number of controlled experiments described in the literature. They argue that none of them makes sense from a methodological point of view. Many researchers have admittedly paid attention to some of the methodological problems involved. According to Kienle and Burkhardt, however, the seriousness of the issue is vastly underrated. Their point is not merely that existing methodology and statistics are often used improperly in medical research, but that the methods we now have - even the most advanced ones - are themselves inadequate in the context of medicine.

An evaluation of the statistics involved is beyond the scope of this book. Kienle and Burkhardt's view is rather extreme, but the points they make must be taken seriously. Notice that their criticism is not an "external" one. Their rejection of orthodox medicine's methodology is not based on an *a priori* commitment to some alternative approach. Kienle and Burkhardt make an effort to work *within* the tradition of natural science and mathematics.

One of the issues raised by them concerns the importance of "individual medical treatment". They argue that common statistics does not provide guidelines for deciding what treatment any particular patient should get. Kienle and Burkhardt value traditional medical judgement reached without controlled experimentation. This makes them share some of the viewpoints now defended by homeopaths such as Mössinger. Ironically, Mössinger brought controlled experimentation to homeopathy.

Recent research on placebo effects confirms the general view defended by Kienle and Burkhardt. A collection of articles edited by White, Tursky and Schwartz (1985) gives an adequate picture of the present state of the art. We will briefly summarize some results and conclusions.

One should notice first that "the placebo effect should be considered as a potent therapeutic intervention rather than a nuisance variable" (Evans, 1985, p. 215). Placebo effects are pervasive, they are by no means limited to chronic disorders which are commonly called psychosomatic.

An accurate assessment of placebo effects, however, is extremely difficult. Indeed the *concept* of placebo is hardly as clear as one would like it to be. Attempts to formulate an explication have uncovered many logical problems (Grünbaum, 1985; Brody, 1985a; see also Grünbaum, 1986a). Moreover, it is far from easy to elaborate correct designs for efficacy studies. The literature is replete with statistical and methodological errors (Rosenthal, 1985).

Other limitations of conventional research on placebo's are even more important. Double-blind studies are not really sufficient! To understand this, one needs to distinguish efficacy research from mechanism research. Wilkins (1985) gives a lucid survey of the problems involved. Suppose a double-blind study shows that some drug results in the disappearance of symptoms associated with some disease. Would that imply that the drug is effective? Yes, it would, but one should realize that this conclusion is valid only if the term "effective" is taken in a very wide sense. It does by no means follow that the effect of the drug is a chemical one. The mechanism may well be psychological. For example, patient interpretations of felt side-effects may cause a change of attitude leading to cure (see also Ney, Collins and Spensor, 1986, who argued that the term "double-blind" is inappropriate in this case). Complicated designs (triple-blind studies, or inclusion of a so-called active placebo control) are needed to take such phenomena into account.

Efficacy research has not been restricted to biomedicine. There is now a great amount of "psychotherapy research" as well. The problems it encounters are partly the same. But psychotherapy is plagued by additional problems (see chapter VI) which have often led to the conclusion that it cannot be approached in a scientific way. This may well be true (if "science" is used in a restrictive sense), but the results of placebo research should inspire a cautious attitude towards therapies based on biology as well. They may not be quite as scientific as the texts of orthodox medicine would suggest.

How are placebo effects to be explained? Various articles in White, Tursky and Schwartz (1985) address this question. Answers are in two categories: there are "mentalistic theories" and "conditioning theories", both supported by recent research. The behavioural approach (cf. conditioning) has shown that humans are not the sole benefactors of placebo effects. Animals respond as well. Biologists should be interested!

We will quote one passage, from Paul (1985, p. 145), which clearly illustrates the import of placebo research. It deals with the situation in psychiatry, but Paul's comments, *mutatis mutandis*, would also apply to other domains of medicine.

Pharmacological investigators, as well as some psychotherapy investigators, have also tended to blur distinctions between the personal-social characteristics of clients or staff and the variables within the class of problem behaviors or psychosocial aspects of treatment techniques, respectively. The result is that conclusions often fail to distinguish justifiable cause-effect relationships from moderating variables or outright confounds. Hundreds of studies, for example, have drawn conclusions about the effectiveness of neuroleptic drugs for treating "schizophrenia" ... on the basis of measured change in "positive symptoms" ... without even assessing the "negative symptoms"... . [Other distinctions are also disregarded; various relevant variables are often not included in efficacy research.] When these additional classes of variables have been included in pharmacological research, conclusions from prior studies have been enhanced, nullified, or reversed, but seldom simply supported ... .

The general flavour of these statements is oddly at variance with the confidence exuded by many texts in biological psychiatry (see chapter V).

Let us try to formulate some general conclusions. Firstly, one can hardly over-estimate our ignorance in matters of health and disease. Even within the domain of orthodox medicine, widely held opinions on biological aspects of health and disease may be mistaken. It is impossible to estimate how much of it makes sense. Methodology often does not get any attention. Studies that take it into account may involve inadequate designs. Methodologically correct research, however, will have to be very complex.

Secondly, the substantive side of medical research is rapidly getting complex. One wonders what would happen if the methodological problems stressed in placebo research would be brought to bear on ordinary research, which often deals with many factors at the same time (cf. analysis of etiology; assessment of physiological and biochemical aspects of disease). The complexity of research designs which are formally adequate *and* cover the factors which are deemed relevant in the average research context almost defies imagination. So it will be necessary to make choices. It is impossible to satisfy, at the same time, all the conditions of adequacy which collectively define good research. Now it is obvious that such conditions are often violated in actual research. Sometimes the ensuing shortcomings are explicitly discussed in research reports, more often they are not. The fundamental limitation of the scientific approach we just formulated is hardly ever mentioned at all. We regard this as a major challenge for the philosophy of medicine.

Thirdly, placebo research should get much more attention in attempts to integrate biology and psychology (cum social science) within medicine. It shows that good research focussing on biology will simply force us to come to grips with "the mental".

It should be obvious by now that one cannot *simply* dismiss current homeopathy for lack of evidence obtained with controlled experimentation. Allopathy and homeopathy have different methodologies, each with its own limitations. Just now, it seems wise not to brandish any option on the basis of a single methodology. We had better concentrate on the development of a more adequate methodology.

Our analysis of homeopathy illustrates that a sensible evaluation of alternative medicines must involve many different issues. In section 7, we will try to integrate the issues we came across. First we will briefly consider another form of alternative medicine, psychic healing, to reveal some new problems.

6. PSYCHIC OR SPIRITUAL HEALING

Psychic or spiritual healers come in many varieties. Some of them claim to effect cures by concentrating on patients even from great distances (for references, see Benor, 1984). What is one to make of this? One would naturally want to know whether the *phenomena* involved in "psychic healing" are real. We will not address this issue, which is highly controversial. An analysis would have to be at least as complicated as our discussion of homeopathy. Here we will concentrate on a different issue which concerns the explanation of phenomena (in section 7 we will comment on relations between the two issues). Should one dismiss claims involving psychic healing at a distance since it cannot be scientifically explained?

The problem with such claims is that psychic healers tend to use terms like "power" or "energy" which for (natural) scientists have special physical connotations and as such have no place in the description of phenomena which seem to violate the laws of physics. Scientists may therefore react as follows. "Healing at a distance is said to involve the transfer of energy from healer to patient. Physical theory describes energy exchange in natural phenomena. No physical theory would appear to account for the postulated transfer of energy. So healing at a distance is a pseudo-phenomenon."

We do not think that this is a valid argument. The healer's premiss is taken at face value without distinguishing between the phenomena in question (healing) and the proffered explanation (power or energy transferred over a distance). One should, of course, concentrate primarily on the phenomenon rather than the theory invoked to explain it. Notice that in psychiatry and psychosomatic medicine, which now belong to orthodox medicine, many well-confirmed phenomena cannot be explained in physical terms. Healing at a distance is a bit stranger than psychotherapy, because it is unclear through what senses information should reach the patient. But one cannot dismiss it as a non-phenomenon simply because the underlying mechanism is unknown. Healing at a distance, and spiritual healing in general, share this lack of explanation with vast areas in orthodox medicine.

Consider the approach of LeShan (1980). After many contacts with psychic healers and people personally acquainted with other kinds of paranormal experience, he elaborated a description of such experiences in general terms. According to him, they typically represent an altered mode of consciousness with four features. Firstly, there is a central perception of the unity of all things. Secondly, time in the ordinary sense is experienced as an illusion. Thirdly, evil and illness are seen as mere appearance. Fourthly, there is awareness of a superior way to obtain information that bypasses the senses. LeShan notes that this description closely resembles descriptions of mystical experience.

LeShan subsequently trained himself, through meditation, to reach the altered mode of consciousness. He claims to have acquired healing powers, and subsequently developed a training program aiming at the development of such powers.

Altered modes of consciousness have a tradition of many centuries. How should one react to this from the perspective of modern science? It is tempting to dismiss any claim of knowledge attained by meditation and related techniques as based on mere subjective perception. Real kowledge, based on science, must have a different, objective nature. But Staal's (1975) perceptive study of mysticism makes shambles of this view. Why should scientific knowledge qualify as the only kind of real knowledge? We will quote Staal's argument at some length.

Not everyone's effort would have the guarantee of a beatic vision and not everyone would obtain *moksa* or *nirvana*; but not everyone would be successful in the Michelson-Morly experiment, or be invited to join the Princeton Institute for advanced studies (p. 58).

A clear candidate for ... a characteristic of mystical doctrines is the distinction between appearance and reality. All mystics assert that there is something real which lies beyond appearances and which is not experienced under normal circumstances. ... Of course, reality presents itself to us as plural, differentiated and, at least partly, embedded in the flow of time. But this does not logically conflict with the possibility that deeper analysis might show that reality is in all these respects different. ... not only is the distinction between appearance and reality consistent with logic and the requirements of rationality; it is in fact exactly what is presupposed in most sciences and all rational inquiry (p. 60).

While mystical doctrines purport to deal with objective reality, mystical experiences have a subjective quality ... . The distinction between appearance and reality is itself applicable to this subjective feature of mystical experiences, sometimes characterized by saying that these experiences are among the 'altered states of consciousness'.

That scholars and scientists are generally suspicious of the 'subjective' is not only due to the difficulty of checking or testing in such cases. There is also some semantic confusion. 'Subjective' means at least two things: (1) 'subjective' as *opposed* to 'objective', where 'objective' means 'objectively true' and 'subjective' therefore means 'false'; (2) 'subjective' in the sense of 'relating to the subject', which is *complemented* by 'objective' in the sense of 'relating to the object'.

The expression 'altered states of consciousness' suggests that the waking state of consciousness, the state for example in which I am when writing and you when reading this, is normal and normative, while mystical and altered states are alterations of it. But of course, here too the appearances might deceive and it is possible that the reverse is true... (p. 62).

Staal's analysis should draw our attention to fundamental metaphysical questions which may arise in the confrontation of orthodox medicine and alternative ways of healing. Orthodox medicine (indeed academic science) has no satisfactory theory of the human mind and no adequate explanation of the connections between the mental and the physical (see Chapters V and VI). This invites extraneous theories about alternative kinds of knowledge and perception.

In section 3, we briefly discussed post-Kuhnian philosophy of science. Kuhn had argued that different "paradigms" may be "incommensurable". A heated discussion ensued. There is now a lot of confusion because central notions such as "paradigm" and "incommensurability" are used in many different senses. We do not want to take sides in current discussions within philosophy. Let us use Kuhn's notions as primitive terms, and take some vagueness for granted.

"Theories" arising from "data" under altered modes of consciousness will obviously not belong to any paradigm of current western science (however the term "paradigm" is interpreted). As yet, one will have to face incommensurability. This suggests that some alternative medicines (e.g. psychic healing) cannot now be brought within the framework of our science. But perhaps current research (in the western tradition) on altered states of consciousness will result in a framework that allows for the assimilation of foreign phenomena.

7. DISCUSSION

Competing schools of medicine, broadly defined, often do not agree about any basic theory. Would it be possible that they reach agreement at the level of phenomena? One would think that facts concerning healing are confirmable somehow, and so will permit some kind of confirmation which is acceptable to opposing parties. Even the claim that all disease is apparent only and healing is the awakening to true reality (a rough translation of the theory of some spiritual healers and faith healers), should involve *some* phenomenal descriptions which are unproblematic: people may fall ill (in the ordinary sense of the term) and may get well again! Is not the real issue about the different theories and their relations to the phenomena? Unfortunately, it isn't.

The trouble is that theories colour the way one looks at phenomena. Moreover, different theories may emphasize very different phenomena. This impedes agreement about significant criteria for healing. Thus there is a big gap between theories of regular medicine and, say, macrobiotic dietary theories, which attach more importance to mental composure and moral rightneousness than to physical symptoms.

Subtle differences between regular medicine and schools which are only "moderately alternative" are at least as noteworthy. Consider the following example. In Western Germany, many old naturopathic traditions now mix with academic medicine to various degrees. The situation has been analysed by Maretzky and Seidler (1985) who point out that individual German physicians, although trained in the same academic environment, align themselves along a continuum between two extremes, "scientific" and "experiential". Naturopathy as experiential medicine ("Erfahrungsmedizin") relies on *experience* as envisaged by the homeopath Mössinger (see section 5.3). It tends to depreciate the objectified symptomatologies of academic medicine: one should not treat diseases but ill persons. Disputes about the efficacy of treatments are thus doomed to remain undecided: "The distinctions in established canons and acceptance of evidence of the divergent medical orientations are cultural in nature" (p. 393).

In practice, consensus is often reached in a negative way. Medical pluriformity becomes an accepted fact, and a "live and let live" compromise is reached after some bickering. Such is the social reality in Western Germany. A similar situation exists in various other countries where local traditional medicine coexists with western academic medicine (e.g. Sri Lanka; see Waxler, 1984), in spite of the promotion of integration by organizations such as WHO (for references concerning WHO programs see Ullrich, 1984).

Similar problems concerning criteria of efficacy even exist within academic medicine. The evaluation of coronary bypass operations is a case in point. Which is the more important criterion, the disappearance or reduction of angina pectoris (perceived by the patient), or the improvement of ventricular function (measured by the cardiologist)? The former criterion makes bypass surgery a great achievement, the latter one still renders it problematic (Rahimtoola, 1982). The issue becomes even more complicated if resumption of physical activities and re-employment are taken into account. Some authors will reject this criterion because too many extrinsic factors are involved, others accept it as they characterize the operation as a *social* failure (in the Netherlands: Dunning, 1980).

Should one resign oneself in the face of persistent disagreement? The question is not easily answered, it is in fact misleading. One will have to distinguish various issues. Are theories underlying medical treatments acceptable? Are the treatments themselves "medically successful"? Are they socially acceptable? Such questions should not be answered by appeal to a *single* set of criteria (for further comments see Kleinman, 1984, who elaborates the issue in the context of cross-cultural comparisons). The disputes we mentioned are often confused as opposing parties are answering different questions. Once that is recognized, at least *some* disagreements may dissolve.

It cannot be denied that regular medicine, from the 16th century onwards, has been successful *with respect to the issues it addressed*. It had its share in the development of natural science and allied technology, which greatly increased predictive and manipulative power. But one should realize what its limitations are if one wants to come to grips with prevailing disagreements over medical treatments.

Firstly, regular medicine has concentrated so much on natural science (especially biology) that it cannot adequately deal with "the mental". This alone should make us cautious in the evaluation of alternative medicines that aim at redressing the balance. Problems concerning "the mental and the physical" will be discussed more fully in chapters V and VI. Secondly, the successes of regular medicine have by no means been as great as its textbooks suggest. If there is but a grain of truth in McKeown's analyses (see chapter II), common medical history may need a lot of rewriting! Thirdly, the methodology of regular medical practice is now facing trouble. Double-blind research, the paradigm of neat efficacy research, has serious shortcomings (see section 5.6). This casts doubts even on the *medical* adequacy of many common treatments.

These critical comments should not be read as a covert defense of alternative medicines. Such medicines have limitations of their own. For one thing, many of them do not have *any* articulated methodology. Instead they seem to rely on common sense. Now we would not like to *exchange* the methodology of science for common sense. Research in psychology has shown that ordinary "experience" can be a dangerous ally.

Would it be possible to develop an integrative theory which combines what is best in regular and in alternative medicine? The answer depends on how the question is interpreted. One will have to specify what one means by "integration". And one will need criteria to decide what is "best".

Let us consider "integration" within the confines of regular science. Philosophy of science, in the first few decades of this century, envisaged a substantively unified science, a grand theory covering sundry disciplines through reduction. Later on, the ideal took a more modest form. Reduction need not be feasible, but one may achieve a unity of method. But even this weaker claim cannot be upheld in the face of recent developments in science and philosophy.

There is indeed much integration within current science. That is, there is a rapidly expanding plethora of connections between disciplines. This kind of integration does not move towards some grand theory. Proliferation of connections between disciplines may give us more disciplines, not less (cf. developments in the biosciences: sexuology, immunology, gerontology, etc.).

In some cases at least, integrations which combine elements from orthodox and alternative medicine should be feasible if the trap of grand-theory-thinking is avoided. But much work remains to be done. Our analysis of homeopathy, which is *relatively* close to regular medicine, shows that differences between regular and alternative medicine are complex. Literature known to us is replete with casual analyses suggesting that such differences can be characterized in simple terms; we can but warn against superficial approaches.

Concerning homeopathy, we have some concrete suggestions which could help to bring it closer to regular medicine. Some of them may also apply to other forms of alternative medicine; homeopathy is just an example we chose to illustrate the spirit of the approach we would favour. For details one should consult the analyses presented in other sections.

General theories of homeopathy (Vithoulkas), at first sight, are not *visibly* testable (whatever philosophical principle of testability one wants to adopt). Items for further research: let homeopaths expound their views of testability; let defenders of orthodox medicine try to *reconstruct* homeopathy so that its theory takes the form of hypotheses they regard as testable.

Theses concerning the effectivity of homeopathic drugs, likewise, call for new approaches. Items for further research: let homeopaths conduct more controlled experiments (cf. Mössinger) to bridge the gap between allopathy and homeopathy; let defenders of orthodox medicine scrutinize the limitations of controlled experimentation (cf. Kienle and Burkhardt; research on placebo's); let anyone concerned (philosopher, statistician, medical researcher) ponder on how one could improve methodology.

CHAPTER IV.   CONCEPTS OF HEALTH AND DISEASE

1. INTRODUCTION

Health is desirable, disease is not. Health and disease are value-laden. Medicine is concerned with health and disease, so it is concerned with values. Medical diagnosis therefore cannot be like, say, biological taxonomy. There is no place for values in taxonomy.

Someone who argues in this way would be called a *normativist* by philosophers of medicine. Normativism has its opponents. There are those who would defend the thesis that medical *science* need not bother about values. Values rather belong to medical practice. *This* view has not been honoured with a generally accepted label. Goosens (1980) calls it *neutralism*, Ruse (1981) prefers the term *naturalism*. We will use Ruse's label.

The argument we started with is concise. Perhaps it does not accurately represent what normativism is about. That is one of the things we want to find out in this chapter.

Should the normativist's argument be accepted? It is not easy to answer this question, there are pitfalls. Consider the first sentence. "Health *is* desirable." What does "is" mean? It may represent a conceptual connection: one could *define* being healthy as being in a state which is desirable, disease representing an undesirable state. This would seem to imply that statements to the effect that somebody is healthy, or has a disease, have a normative component, because definitions of "health" and "disease" mention values. So medical theory, as far as it needs these concepts, would not be wholly empirical. Values would get a place in medicine which they don't have in empirical sciences like biology. Alternatively, "health" and "disease" could be defined in value-neutral terms. In that case the phrase that health is desirable would express an *opinion*. The argument would then collapse. Lots of things which one investigates are desirable or undesirable in a non-definitional way. There is much research in biology on locust swarms in agricultural areas. That does not imply that biologists, *qua* biologists, need to bother about their being (un)desirable. The desirability of health could be like that.

Normativists apparently think that the "is" in "health is desirable" represents a definitional connection. Otherwise they would only be stating the obvious, viz. that physicians are somehow concerned with values. Normativism seems to defend the additional, more specific thesis that values colour the meanings of central concepts in medicine. This seems to suggest that medicine cannot have "real" science, or that medical science is special.

The controversy between naturalism and normativism will be discussed in sections 2-5. We intend to show that the views of Boorse, whose naturalism is well-known, must be rejected. However, we will criticize theses of normativism as well. Sections 2 and 3 introduce arguments of some representatives of the two views and preliminary comments. We will subsequently show, in sections 4 and 5, that the terms in which the controversy is usually stated are misleading. The opposing parties seem to be quarreling about words. But the real stakes are higher. They involve complex relations between theory and practice in medicine which one cannot clarify with mere conceptual analysis.

2. HEALTH AND DISEASE IN A BIOLOGICAL PERSPECTIVE

2.1. Against normativism: Boorse

Boorse (1975, 1976, 1977), in his attack on normativism, is concerned with three interrelated questions. Firstly, can mental health and mental illness be subsumed under the concepts of health and illness used in somatic medicine? Secondly, what is the relation between illness and disease? Thirdly, should health and disease be defined in a value-free way? We will concentrate on the third question. Boorse's way of construing a positive answer to this question is summarized below.

In clinical contexts, disease is synonymous with "unhealthy condition" and health represents normality (Boorse, 1975, p. 50). But normality in a simple statistical sense seems neither necessary nor sufficient for clinical normality. Many authors have argued therefore that medical judgements contain normative components. They associate health with notions like goodness, desirability and approval, and disease with badness, undesirability and disapproval. This kind of normativism is conspicuous in psychiatric texts (for quotations see Boorse, 1975 and 1976), but it affects somatic medicine as well. Somatic diseases are often regarded as undesirable conditions that doctors happen to treat. "As medical practice varies over time with evolving social institutions and values, so will the inventory of unhealthy conditions" (Boorse, 1977, p. 545).

Boorse admits that values and customs play a part in the discourse of medical practice. His objections to normativism concern medical theory, which allegedly has an entirely different conceptual framework. The following quotations illustrate his line of reasoning.

[Medical practice does not help us to define disease because] doctors also treat some conditions they do not regard as disease. Among standard medical procedures are circumcision, cosmetic surgery, elective abortions and the prescription of contraceptives. None of the conditions so altered appears in the AMA Standard Nomenclature, the latest attempt at a comprehensive listing of diseases. Nor are they listed as diseases by other medical texts (Boorse, 1977, pp. 545-546).

[Pain, suffering and discomfort are sometimes mentioned as markers of disease.] This idea suggests a focus on medical practice rather than theory, and in fact on patients who come complaining of symptoms. Even within medical practice, routine physicals can disclose asymptomatic disease ... . As textbooks of medicine constantly mention, a complete absence of 'subjective distress', is compatible with severe internal lesions (ibid., p. 547).

Unlike chemists or astronomers, physicians and psychotherapists are professionally engaged in practical judgements about how certain people ought to be treated. It would not be surprising if the terms in which such practical judgements are formulated have normative content. ... But behind this conceptual framework of medical practice stands an autonomous framework of medical theory, a body of doctrine that describes the functioning of a healthy body, classifies various deviations from such functioning as diseases, predicts their behaviour under various forms of treatment, etc. This theoretical corpus looks in every way continuous with theory in biology and the other natural sciences, and I believe it to be value-free (Boorse, 1975, pp. 55-56).

The difference between the two frameworks, of practice and of theory, is underlined by associating negative evaluations of diseases with the term "illness" of medical practice, and reserving a neutrally descriptive term, "disease", for medical theory. "The point is that illnesses are merely a subclass of diseases, namely, those diseases that have certain normative features reflected in the institutions of medical practice" (ibid., p. 56). Boorse then argues that an illness must be a reasonably serious disease with incapacitating effects, and this requires that it be (1) undesirable for its bearer, (2) a title to special treatment, and (3) a valid excuse for behaviour that we criticize in other situations (Boorse, 1976, p. 63).

Boorse then elaborates "health" and "disease" as value-free descriptive concepts within medical theory. His 1977 paper gives the most details. He initially characterizes health as "normal functioning, where the normality is statistical and the functions biological" (p. 542). The following definitions capture this idea in more accurate terms.

1. The *reference class* is a natural class of organisms of uniform functional design; specifically an age group of a sex of a species.
2. A *normal function* of a part or process within members of the reference class is a statistically typical contribution by it to their individual survival and reproduction.
3. *Health* in a member of the reference class is *normal functional ability*: the readiness of each internal part to perform all its normal functions on typical occasions with at least typical efficiency.
4. A *disease* is a type of internal state which impairs health, i.e. reduces one or more functional abilities below typical efficiency (Boorse, 1977, p. 555).

Boorse justifies his definitions in an extensive account which contains the following elements. Functions in biology are related to a hierarchy of goals, the highest level being to some extent indeterminate (it "must be determined by a biologist's interest"; ibid., p. 556). In the context of medicine, however, we need not bother about goals being indeterminate. "But it is only the subfield of physiology whose functions seem relevant to health. On the basis of what appears in physiology texts, I suggest that these functions are, specifically, contributions to individual survival and reproduction" (p. 556). Anyhow, "whatever goals are chosen, function statements will be value-free, since what makes a causal contribution to a biological goal is certainly an empirical matter" (p. 556).

"Typical" traits or "standard" performances relative to a reference class play an important role in the definitions. It is obvious that they involve idealization:

... the subject matter of comparative physiology is a series of ideal types of organisms .... [However,] the idealization is of course statistical, not moral or esthetic or normative in any other way. For each type a textbook provides a composite portrait of what I will call the *species design*, i.e. the typical hierarchy of interlocking functional systems that support the life of organisms of that type. ... The species design that emerges is an empirical ideal which, I suggest, serves as the basis for health judgements in any species where we make such judgements (p. 557).

Idealizations, as Boorse views them, represent a *typological* approach, a kind of Aristotelian biology (p. 554). Boorse admits that such an approach has its limitations, but he argues that it is appropriate for dealing with health and disease. For example, the notion of species design is inappropriate on evolutionary time scales, but this does not bother him. Medicine relies on the short-term constancy of the human species.

On all but evolutionary time scales, biological designs have a massive constancy vigorously maintained by normalizing selection. It is this short-term constancy on which the theory and practice of medicine rely. ... Our species and others are in fact highly uniform in structure and function; otherwise there would be no point to the extreme detail in textbooks of human physiology (p. 557).

Boorse recognizes two "anomalies" that do not fit his definitions, structural disorders and "universal diseases". Purely structural disorders listed in the AMA Nomenclature without reference to functional impairment, however, may have been included on the assumption that the affected organ has an unknown function (cf. congenital absence of the appendix, calcification of the pineal gland). "But the Nomenclature also lists minor deformities, especially those of the nose, the ear, and, mysteriously, the hymen. Many of these deformities disturb no normal function ..." (p. 565). Perhaps they have been listed "for convenience in record-keeping". "Any structural disorders that do not fit these two categories remain anomalies in an otherwise intelligible scheme of classification" (p. 566).

Boorse gives more weight to "universal diseases" as an anomaly. Within a given reference class, disorders may be statistically normal (e.g., dental caries, atherosclerosis and benign hypertrophy of the prostate). If that is the case, "it is clear that medicine is prepared to view the entire reference class as functioning abnormally" (p. 566). Part of this anomaly (relative to Boorse's definitions) is removed by the clause that "normal functioning" implies that every body part functions in its typical way. When "universal diseases" affect different members of the reference class in a different way, as they often do, they are in fact diseases on Boorse's account; "... the only problem arises when everybody has the same disease in the same location" (p. 566). This seems to apply only to universal diseases "which are evenly distributed, e.g. lung irritation due to environmental pollution or arterial thickening after a certain age" (p. 566).

Boorse's way out is to suggest that such disorders are not related to the species design. That is, they are caused by environmental agents rather than genetic factors. For this reason, he expands his definition of health to accomodate "environmental injuries" as diseases.

A disease is a type of internal state which is either an impairment of normal functional ability, i.e. a reduction of one or more functional abilities below typical efficiency, or a limitation on functional ability caused by environmental agents. [This view of disease permits Boorse to endorse the traditional, simple definition of health:] health is the absence of disease (p. 567).

The definitions have important consequences. "What there cannot be, on this view, is a universal genetic disease". Boorse consequently refuses to regard senile decline of function as a disease. It is a puzzle "... why old age is not always seen as a stage with its own statistical norms of healthy functioning" (p. 567).

Boorse regards normativism as represented by Engelhardt (allegedly the main alternative to his approach) as singularly unattractive. It implies that the concept of disease has no exact content because it is a vehicle for changing human goals and expectations. These ideas "... do little to explain the actual medical inventory of diseases. Such accounts cannot explain this inventory because they cannot predict it" (p. 567). In this connection, a final point in Boorse's struggle to eliminate values is important, *viz.* his discussion of the concept of "positive health".

The proposition that health is the absence of disease does strongly contrasts with the famous WHO definition which invokes the ideals of complete physical, mental and social well-being. Such ideals look commendable, but they are rather impractical. First and foremost, they change health from a limited to an unlimited goal. There are no clear limits to functional excellence: how strong can humans become, or how intelligent? Moreover, we will have to face incompatibility of specific goals. It is impossible simultaneously to maximize one's performance in weightlifting and in playing the piano. Any assessment of positive health will therefore necessitate the weighing of values.

This value-ladenness is the most striking difference between positive health and the traditional negative variety. Our conception of disease required no value judgement about what forms of human life are admirable or desirable. Diseases were interferences with an empirically discoverable species design. Thus what it is to eliminate disease is uniquely describable in advance of normative decisions (p. 571).

We have now covered the essentials of Boorse's arguments. Does his view of health and disease make sense? It has a plausible ring. But as shown in the next section, his reasoning is just too casual.

2.2 *Towards a sensible biology: Boorse's views refuted*

Empirical *versus* normative medicine. That distinction is Boorse's main subject. He opts for an empirical approach of medical *theory*. One should notice that he introduces many other distinctions which colour his version of naturalism.

Firstly, Boorse chooses to make biology the paradigm of scientific medicine in his discussion of health and disease. But are psychology and social science really irrelevant? (Chapter VI will deal with this subject.) Secondly, Boorse decides to concentrate on a limited area of biology, viz. physiology. Thirdly, his physiology is of a special kind. Normal physiological processes, in Boorse's view, belong to the species design. And his conception of the species design involves yet another distinction, between internal (genetic) causes and environmental causes of disease. Fourthly, Boorse favours a typological approach of species designs, i.e. he gives biological variability a minor role. Characters of organisms (in the present context: functional abilities) are supposed to be distinct and largely uniform within species or reference groups (within species) such as sexes. Thus it should be possible to elaborate neat characterizations of species. Boorse's comments on pathology suggest that he also favours a typological approach of disease classification. That is not surprising since he wants to define diseases in terms of biological characters.

Boorse's arguments draw on all these items at the same time. For the sake of clarity, we want to keep them apart in our comments. We will first argue that Boorse's description of medicine is implausible. It represents a typological way of thinking which is foreign to actual medicine. But the main point of criticism, which we will subsequently elaborate, is that Boorse's typological approach represents bad biology. For the record, other comments on his naturalism need to be mentioned here. Goosens (1980) has convincingly argued that Boorse has attacked a very strong form of normativism which indeed is commonly defended in the literature. A weaker form would remain unscathed under his attack. Brown (1985a) gives an overview of technical problems with Boorse's definitions.

Is it possible to develop an "objective" view of health and disease (and *particular* diseases) on the basis of medical theory? Boorse is fairly positive here, too positive we think. His picture of medical theory is rather biased because he disregards the history of medicine, and because he relies on texts such as the AMA Nomenclature. His discussion of "universal diseases", in which the distinction of internal and environmental causes plays a crucial role, shows how he simplifies medicine.

Consider how medicine actually deals with the notion that functional disabilities may be ascribed to internal or to environmental causes. (We will discuss the *merits* of this distinction as a separate issue.) Medical thinking quite often vacillates between various types of causation. For example, the discovery of new pathogenic agents (cf. *Legionella pneumophila*) may start a phase of purely monocausal thinking, which is followed after a while by the recognition that the agent is apparently ubiquitous. Some constitutional explanation is then developed to account for the low frequency of overt clinical disease. If it fails in a particular case, other environmental factors may get the blame. And so on. Such cycles are common. They show that one cannot approach disease without being selective. *An inventory of all the factors that influence a disease is not possible if only for practical reasons.* Now one could argue that some factors should have a privileged position for *theoretical* reasons but, as we will argue below, this would involve a commitment to unacceptable typology.

The distinction of natural ageing *versus* age-related disease, likewise, is not as neat as Boorse suggests. Research on age-related dysfunctions continually generates distinct diseases where medicine formerly recognized one homogeneous phenomenon of natural ageing. In the early 19th century, most chronic joint diseases, including the trophic deformities caused by tertiary syphilis, were lumped together as natural effects of wear and tear. Up to some 40 years ago, the various forms of arteriosclerosis and arterial atheromatosis were similarly considered to be a syndrome of natural ageing. Senile dementia is a more recent case in point. The differentiation between Alzheimer disease-type dementia and multiple infarction or arteriosclerotic dementia of the aged dates from the early sixties.

These developments so far have resulted in continually shifting boundaries between age-related diseases and "normal" ageing, however defined. Many conditions which are now known as diseases, e.g. rheumatoid arthritis, arthrosis deformans, arteriosclerosis and Alzheimer's disease, were at one time all included in normal ageing under the usual gerontological criteria of universality, deleteriousness, progressivity and irreversibility. And Boorse's criteria of age-related normality and internal causation would in the past have led to the same result. In general, pathological changes accumulating in old age can be found, more sparsely, at younger ages too. Often a genetical predisposition with variable latency periods is involved (e.g. diabetes, hypertension). Interaction of genetical and environmental factors, differentially proportioned, are a better explanation than any concept of "normal" ageing (cf. Thung, 1957, 1962, 1964, 1965).

Could there be a stable list of diseases in the future? We do not think so. Possibilities for making new distinctions in the study of functions do not seem to have natural limits. Boorse at any rate does not give arguments that make the existence of such limits plausible. History shows that there has not been any steady progress towards a fixed array of diseases. Authors who are aware of history's vicissitudes, however, have nevertheless defended views that closely resemble those of Boorse. Daniels (1985) even thinks that the fallibility of medical judgements revealed by history, should count in favour of Boorse's naturalism. By way of an example, he argues that (for normative reasons) some conditions or behaviours (e.g. masturbation) may at one time have been *viewed* as diseases. But this does not make them diseases. *We know better.*

This is a rather optimistic view. Modern nosologies may be superior to older ones in many respects. But it does not follow that there are now stable and value-free distinctions.

Boorse's views on *biology* are even more problematic. His typological approach of species designs would be rejected by almost all biologists. In his 1977 paper he briefly defends typology as an Aristotelian approach which may still be adequate in some contexts. Physiology, especially its textbooks, is *his* context. However, physiology books are not a very good source for information on species designs. The subject belongs primarily to taxonomy, and taxonomists, to whatever school they may belong, nowadays emphatically reject the typological way of thinking (see e.g. Mayr, 1969, 1982; Sneath and Sokal, 1973). Philosophers concentrating on taxonomy share their opinion. For example, David Hull, a well-known philosopher of biology, wrote a paper in 1965 with the ominous title "The effect of essentialism on taxonomy - two thousand years of stasis" ("essentialism", in the present context, roughly means "typology"). One of his points is that species names cannot be defined in terms of necessary and sufficient properties (see also Beckner, 1968). Variability within species simply does not allow that, even if we concentrate on species here and now and, like Boorse, disregard evolutionary time-scales. The point has been granted by taxonomists. There is much discussion about species now, but the issue of typology has led to consensus.

Variability within species is also an important theme in population genetics. Researchers in this field agree that the amount of genetic variation within species and populations is enormous (Lewontin, 1974). We will give one specific example to show how Boorse distorts biology. The example will also illuminate our next theme, internal (genetic) *versus* environmental causation.

Aboriginals of some tropical areas are plagued by a crippling disease, sickle cell anemia. The name is almost self-explanatory. People who suffer from the disease have red blood cells which easily become deformed, sickle-shaped, by cristallization of an abnormal hemoglobin (HbS). This results in hemolysis with chronic anemia, thrombosis, and other complications. One would expect that natural selection will tend to eliminate the disease, but this has not happened. How can one explain the rather stable incidence of the disease in some populations? The sickle cells only appear spontaneously in people who are homozygous for the gene involved in the disease. Heterozygous people, who have one copy of the gene beside one copy of a "normal" gene, are usually clinically healthy because their blood contains only 30-40% of abnormal hemoglobin. *These* people have an unexpected advantage over others who are homozygous for the normal gene. They will not easily get the tropical form of malaria caused by *Plasmodium falciparum*. Selection obviously fosters a compromise by keeping sickle cell anemia (at a low frequency) in the population. Compromises of this kind are very common. They explain genetic variability (at least partly), and they also illuminate the interplay of genetic and environmental factors.

Now Boorse could object to our analysis that he does account for variable characters. The sickle cell gene is much rarer than the normal gene in populations where the disease occurs. So it does not belong to the species design. Likewise for other variable features. The most common, "normal" variants represent the species design.

However, this would imply that all individuals would fall short of the design of their species. Variability is so pervasive that everyone will have *some* characters and functional abilities that are "abnormal" on Boorse's view. True, Boorse admits that designs represent *ideals*. But why should one make ideals which biology rejects a paradigm? Biologists would argue that variability is part and parcel of species designs.

Boorse suggests that genetically programmed species designs are somehow invariant although the environment may *happen* to interfere with them. But the association of genetic and environmental factors is much more intimate than that. *All* characters of organisms depend on the joint effects of genes and the environment. So all of them are "genetically determined" *and* "environmentally determined". Statements to the effect that some character is genetically determined may be sensible if one recognizes that they represent an elliptical way of speaking. They should refer to genetic factors which affect the character in the same way in all the environments one is interested in. For that matter, most characters are not determined in this way. Neither are they "determined by the environment".

Many authors in medicine, and in the philosophy of medicine, have misconstrued the distinction of genetic and environmental causation. We have stressed the issue since Boorse's views are by no means an isolated example. Chapter VI will show that much research in psychiatry, and psychosomatic medicine, even associates the distinction with a rather problematic demarcation of disciplines. Other examples were already given in chapter II (section 4.3) and chapter III (section 2.1).

Boorse's conception of design becomes inadequate even within physiology, his favourite discipline, as one leaves the realm of elementary textbooks of medical physiology. Physiological functions change with the environment, so *there are no reference values simpliciter*. Reference-values will have to be *context-dependent*. They are sensible only if they are related to the environment besides age and sex. Blood cell counts change with altitude, metabolic rates with temperature, and so on. Clinical diagnosis in fact makes allowance for such items. Many conditions which are pathological according to Boorse's definitions may be biologically normal in some, and abnormal in other environments. Plain biology would not justify their being classified as diseases in our nomenclatures.

Boorse apparently wants to use idealizations representing optimal designs in hypothetical environments without pathogens and harsh physical conditions. Such environments are not very "biological". They represent ideals one would like to realize. Boorse's idealizations are nonsense in the context of biology. But they may have a function if we regard them as a means for expressing human values. On that interpretation Boorse is just another normativist.

The arguments we gave were not meant to demonstrate that there cannot be any viable variant of naturalism. Boorse's variant, however, does not qualify. It could develop into a sensible form of naturalism only after wholesale reconstruction.

3. THE PHILOSOPHY OF NORMATIVISM

3.1. Preliminaries

Boorse's primary aim was to keep values out of medical theory in order to give medicine a bit of real science. Our comments in the previous section showed that he did not succeed since he put a rather distorted kind of *science* at the basis of medical theory. This, of course, is not a sufficient reason to accept normativism as an alternative to Boorse's naturalism. So we need to analyse the arguments which normativists themselves have offered in favour of their position.

Medicine, as viewed by normativism, cannot be value-free. Even its theory is concerned with values as the *concepts* of health and disease cannot be defined without reference to values.

Normativism has our sympathy, but we will show that some positions defended by its adherents are *philosophically* inadequate. By way of a starting point, we will discuss contributions of Engelhardt and Gräsbeck to a symposium held in Sweden in 1982 (session on concepts of health and disease), published in Nordenfelt and Lindahl (1984).

3.2. The clinical perspective: Engelhardt

Engelhardt (1984), in his defense of normativism, does not try to elaborate general concepts of health and disease. Instead, he concentrates on medical practice which needs an inventory of clinical problems (a quite heterogeneous set) rather than a catalogue of "diseases".

Engelhardt first puts philosophical and conceptual issues in a historical perspective. Throughout the centuries there have been disputes "with regard to whether diseases have a reality in and of themselves, or whether they are the creations of the physicians" (p. 27). The two views are known as the *ontological* account and the *physiological* account, respectively. (Notice that "physiological" is not used in the ordinary sense.)

The disputes may be construed either as arguments over the status of disease classifications (natural *versus* artificial), or as arguments over useful ways to approach medical reality. Engelhardt favours the latter interpretation in view of medicine's therapeutic goals; "... an instrumentalist understanding of medical generalization would appear to be in accord with the way in which medicine is usually practiced" (p. 28).

Engelhardt subsequently discusses a second controversy concerning concepts of disease. Are the concepts value-dependent? He formulates various objections against Boorse's view that they are not.

The main difficulty with Boorse's position is that it will either describe a state of affairs of interest to taxonomists classifying the peculiarities of particular species, and not be equivalent to the concepts of disease employed by physicians, or it will indeed require an appeal to particular values (p. 29).

In fact, if one examines the problems of defining health within a post-Darwinian understanding of biology, one sees that it will not be possible in fact to have one sense of health or of disease. Instead, one will find various families of concepts. If one attempts to understand health as a successful adaptation to a particular environment, one will need to specify not only the environment, but the goals for success. ... A dimension of such an environment will be the culture in which the humans in question live (p. 30).

And goals will have to be specified. One needs to know "ranges of goods in which individuals tend to have interest", and one should be "interested in determining which goals medicine can aid in achieving. Thus, disease concepts will involve reference to problems in achieving expected abilities to function, freedom from pain, and expected human grace and form, inso-far as these problems are seen to be physiologically or psychologically based and beyond the direct and immediate will of the individual whose problem it is " (p. 30).

From the standpoint of clinical medicine, "the search for species-typical deviations of physiological or anatomical functions or structure often serves as a good heuristic device" (p. 32) in the characterization of diseases. But this will not suffice. Goals of patients and social setting will also have to play a role. So values will enter the picture.

Evaluations play different roles:

Clusters of findings exist as explananda for medicine because they cause states of affairs held in some sense to be improper. ... Patho-physiology is distinguished from physiology as being the physiology of some variety of suffering. [Further, the explanations we give will depend on the context, even on economic considerations]. Thus, tuberculosis is likely to be characterized as an infectious disease by internists, even though individuals working in social medicine may see it as a disease tied to poverty, or those working in genetics as a disease that may have important hereditary components (p. 32).

Engelhardt grants that different concepts of disease may be used outside the clinical context.

Here one should note that the meaning of scientific terms is best found in the processes in which they are employed. As a consequence, one will need to distinguish between concepts of disease as they exist in unapplied "medical" sciences versus applied medical sciences. ... In the first case one finds "diseases" as elements of a process of explanation. In the second case one finds "diseases" as elements of a process of intervention (pp. 34-35).

[Clinical medicine is an applied science,] it is a search for explanations and predictions in the service of non-epistemic goals such as the achievement of well-being or the avoidance of impairments. Since one is interested in applying knowledge to non-epistemic goals, the success of such applications is judged primarily by non-epistemic standards (p. 35).

The nature of clinical medicine is also revealed by its notions of causation. Clinical problems are mostly affected by many factors, which will not equally interest us:

The language of diseases, or better still, clinical problems, is ... placed within a pragmatic account of causation in which those factors which are usually held to be accountable in therapeutic practices are held to be *the* causes. ... Thus, phenolketonuria is likely to be seen as a genetic clinical problem or a metabolic clinical problem, depending on whether one is providing genetic counseling to prospective parents who carry the recessive genes for the trait, or whether one is explaining to parents the reasons for providing a phenylalanine-free diet to their newborn child (pp. 35-36).

Clinical theory, in short, will not support any reification of disease entities. "Talking of "clinical problems" rather than "diseases" can ... function as a linguistic reformulation aimed at freeing analyses from old metaphysical associations" (pp. 36-37).

Engelhardt's plea for normativism is cogent. We agree with many of the points he is making, but we want to change emphases. To begin with, consider how Engelhardt reasons in elaborating an alternative to Boorse's views. The argument roughly goes as follows. "An adequate (non-Boorsian) biological view of health and disease will have to consider the environment. Culture is a dimension of the environment, and culture involves human values. So we will have to characterize "health" and "disease" (if we want to use such concepts at all) in terms of human values". We have no problems with this line of reasoning if it is meant to show that medicine is concerned with values as it deals with health and disease. But it need not follow that the *concepts* of health and disease must refer to values. We do not know whether Engelhardt would wish to use concepts in this way since his formulations are somewhat ambiguous (see also section 4.1). Anyway, our own view is that it may be useful to characterize biological aspects of conditions we regard as "diseases" with separate concepts. True, biological

factors, psychological factors, culture, human values, etc., all play a role in
health and disease. But it does not follow that *concepts* of health and disease
must cover all and sundry items. Specifically, it may not be necessary to
define the concepts in terms of values even though health and disease
involve values. The point is by no means trivial since it affects the status of
science in medicine.

Even if concepts of health and disease *are* defined in terms of values,
theories of (clinical) medicine may well remain empirical (see Macklin,
1981, pp. 415-416; Munson, 1981, p. 202). Consider the statement that
health is desirable. We have already argued in section 1 that "is" in this
statement may either represent a conceptual connection or a factual
connection. Perhaps normativists like Engelhardt would opt for the first
alternative by *defining* health in terms of desirability (or other terms
connoting values). At first sight, this would imply that expressions like "X
is healthy" (or "X has a disease") have normative force; we did take this for
granted in section 1. However, they can still be interpreted as nonnormative, empirical claims. The point is that definitions in terms of
desirability can be read in different ways. "Being desirable" could be taken
to *express* a value. But it could also be used to describe what people do *in
fact* desire. The normativist thesis that medicine must deal with values,
anyway admits of very different interpretations. It could refer to medicine
as a normative enterprise. But there is also a weaker interpretation which
allows medicine, at least medical theory, to remain empirical. We do not
know whether Engelhardt would wish to defend the stronger option.

Engelhardt rightly stresses the importance of the *context* for
explications of "health" and "disease". But we do not think that context-
dependency makes clinical medicine special. In plain biological research
one also has to face the problem of selecting relevant factors to be treated as
"the" causes of some phenomenon. A great variety of contextual factors
will influence selections, just like in clinical medicine. Moreover, Engelhardt wrongly belittles the role of science in clinical medicine. Consider his
distinction of unapplied medical sciences and applied medical sciences (cf.
quotation from pp. 34-35). The second category, according to him, is very
different from the first one because it is *applied* science. We would like to
change the emphasis. The two categories have much in common since the
second category is applied *science*. Engelhardt pays so much attention to the
themes of values and context in his discussion of clinical medicine, that the
role of science becomes obscure. He rightly notes that unapplied medical
science is concerned with nosologies, and applied medical science with

clinical problems. But he almost seems to forget that nosologies will have to play *some* role in the solution of clinical problems. Thus the relations between medical theory and medical practice remain unclear. This issue will be taken up again in section 5.

3.3. *Normative laboratory medicine: Gräsbeck*

Gräsbeck (1984), like Engelhardt, puts clinical medicine in the perspective of normativism. His subject is health and disease from the point of view of the clinical laboratory. Even in the clean setting of the laboratory, supposedly a source of purely objective and quantitative data, there are no unambiguous and objective criteria of health or normality.

[Laboratory tests are ordered] not to distinguish between health and disease but one disease from another. ... Even so, reference values from persons who are healthy are of great importance. One way of deciding that the patient is getting better and that the treatment is efficient is to observe that his laboratory tests give results similar to those of healthy persons. In laboratory medicine (as in medicine in general) health represents the goal and health-associated reference values are goal values (p. 48).

So one needs goal-oriented definitions of "health". The WHO definition is not very useful; even the WHO itself does not take it seriously. Gräsbeck mentions two "realistic" alternatives of his own.

1. Health is characterised by a minimum of subjective feelings and objective signs of disease, assessed in relation to the social situation of the subject and the purpose of the medical activity, and is in the absolute sense an unattainable ideal state. 2. Health is an abstraction representing the chief goal of medicine. Definition 1 expresses current views among biomedical scientists. Definition 2 is aphorism-like and attempts to express the current use of the word "health" by politicians and administrators (p. 49).

How to assess health in actual practice? How to use laboratory tests? Before discussing such questions, Gräsbeck briefly mentions two dichotomies that point to different conceptions of health, holistic *versus* non-holistic health and subjective *versus* objective health. He argues that non-holistic, objective health is the target of laboratory tests.

It is not easy to develop objective laboratory criteria. The use of reference values representing entire populations is not very sensible. Of course one can work with age-related reference values instead of the classical "normal values", but age is hardly the only confounding factor. There are many sources of variation, and intra-individual variation must be set apart from inter-individual variation.

The arbitrariness of classical reference values is also revealed by the use of "risk-levels" (e.g. serum lipid concentrations or blood pressure levels which predispose to diseases of the circulatory system).

... there is no dichotomy between values associated with high and low risk, but there is a smooth change in the risk level as the values of laboratory tests increase. However, one can arbitrarily decide upon a risk limit and draw a reference limit below (or sometimes above) which the values are statistically associated with a low or moderate risk. Such values are even more goal-oriented than the ordinary health-associated reference values (p. 51).

More fundamentally, the elaboration of objective standards for health is hampered by the pervasiveness of disease. The number of genes involved in diseases is very large, and there is much heterozygotism. It is probable therefore that "nobody is completely healthy, ... but as long as we live, we possess some health" (p. 53). In practice, if we try to obtain healthy controls, our samples will become smaller and smaller as we try to eliminate subjects, e.g. because they are taking drugs (aspirin, contraceptives, alcohol) which influence serum components, or because they have some minor ailment. Nearly nobody is "normal" for all functional variables.

As to disease, there are typical values for laboratory data associated with certain diseases, but the curves describing the values in typically healthy and diseased populations always overlap. In individual cases, diagnosis is very subjective. Diagnoses are actually confirmed or rejected afterwards on the basis of therapeutic success or failure.

Even the choice of tests, and of labels for diseases, may be quite arbitrary. Consider tuberculosis. Traditionally, it was first and foremost regarded as an infectious disease caused by the tubercle bacillus. Special drugs were developed and special legislation was passed to deal with tuberculosis infections. Today, however, more risk factors are recognized: old age, poor hygiene, starvation, alcoholism. Persons who happen to get the disease might just as well have had a different infection. In developed countries, tuberculosis is in fact almost regarded now as just another case of infectious lung disease. The basic condition might be labelled immunological insufficiency.

Gräsbeck concludes that laboratory tests do support the clinician in his classificatory work. Individual clinical problems have to be categorized for practical reasons. Standard patterns of dealing with patients are needed but they are arbitrary to a large extent, and they rapidly change with time. Laboratory medicine continually influences the nosologies of clinicians. So it is to be expected that the taxonomy of diseases in the near future will be very different from that utilized today.

Gräsbeck ends with an optimistic note which seems to contradict the spirit of his paper. Present medicine is based on a mixture of tradition and practical experience, and theoretical and experimental considerations. Much of it is still an "art" rather than a science, but "the scientific way of thinking is steadily seeping upwards from' the experimental biomedical sciences into clinical medicine" (p. 58). Laboratory medicine is closest to the basic sciences and it is quickly adopting strict definitions, statistical ways of thinking and appropriate controls. "This attitude is also reflected in the way we conceive health and disease" (p. 58).

The problems discussed by Gräsbeck are typically problems for "clinical decision making", a modern discipline which he does not discuss. The remarks at the end of his paper could have been made by a representative of this discipline. Before commenting on Gräsbeck's views we will therefore give a brief description of decision theory in medicine.

Clinical decision making is traceable to Feinstein's original work in the early sixties, on disease taxonomy, which resulted in his "Clinical judgement" (1967). Aware of the many subjective elements in clinical diagnosis and the high interobserver variability of clinical data, Feinstein tried to develop more objective criteria for auscultation and other diagnostic observations. His aim was to uncover the scientific rationality hiding behind the element of "art" in medicine and to strengthen it. The massive growth of diagnostic and therapeutic technology swamps the clinician with information which defies his cognitive abilities. Feinstein proposes new, more explicit ways of treating this information. Set theory and Venn diagrams arc among his tools.

This project did improve the reliability and the rational handling of diagnostic data. But it also resulted in a detailed splitting of diseases into subgroups with overlapping signs and symptoms. This necessitated a new approach to diagnostic taxonomy and ultimately led to a reconstruction of the entire process of medical judgement.

Feinstein's work was not an isolated phenomenon. Several authors developed similar ideas in the same period (e.g. Lusted, 1968). Nearly a decade later, Wulff (1976) wrote his now classical book on "rational diagnosis and treatment", which summarizes and expands early efforts to make medical practice rational (see also Wulff, 1986, and Feinstein, 1985, 1987, for recent comments). By then, no place was left for diseases as discrete phenomena. We rather have a continuum along many dimensions.

Clinical decision making has subsequently developed into a discipline in its own right, which aims at strictly rational rules and quantitative methods for diagnostic and therapeutic decisions (cf. Weinstein and Fineberg, 1980).

Except for the final paragraph, the atmosphere of Gräsbeck's article is clearly at odds with the spirit of medical decision theory. He almost suggests that there is not really a rational way to use laboratory data in clinical medicine. Representatives of medical decision theory have a different attitude. Theirs is the optimism of a new rationality. How can one explain the difference?

Perhaps it is merely a matter of emphasis. Decision theory is a theory, but for medicine it functions as a *method*. Once one has reference values, goals, data deemed relevant, etc., decision theory may work in a perfectly rational way. Medical decision theorists often take these items for granted as they concentrate on methods. Gräsbeck scrutinizes the requisites which form the context in which the methods are used. And the problems lie in this context.

Ideally, one should have good *medical* theories, and a sensible view of the goals and constraints of medical practice, in order to use decision theory in the context of medicine. Once one realizes this, it is obvious that decision theory is not an appropriate tool for dissolving the controversy of naturalism *versus* normativism.

4. TOWARDS A NEW RESEARCH PROGRAM

4.1. Logical puzzles and their consequences

The discussion of naturalism and normativism in sections 2 and 3 has uncovered problems which need to be put in a general perspective. We will do this in section 4.2. The present section deals with various issues which naturalists and normativists have hardly considered. To begin with, some general puzzles concerning concept formation are considered.

So far, we have presupposed that it is possible to distinguish definitional connections from other relations. Naturalists and normativists also seem to assume that this is possible, otherwise all the talk about definitions would be unintelligible. Unfortunately, modern philosophy of science has shown that this assumption is problematic. The philosopher Quine (1953, 1960), for example, has argued that our concepts form an interconnected whole, so that we cannot isolate the meaning of individual concepts. As a result, there would not be any sharp distinction between logical (definitional) and factual matters. Quine's *holism* of meaning is not generally accepted (for comments, see e.g. Hacking, 1983). *No* theory of meaning is.

The distinction of facts and values, likewise, has recently been called into question, notably by the philosophers Putnam (1981) and MacIntyre (1981). So we must also face a fact-value holism. An inspection of some quotations from papers by normativists will show how such holisms are related to our present theme.

Engelhardt (1981) characterized the role of values in medicine as follows.

...evaluation enters into the enterprise of medical explanation because accounts of disease are immediately focussed on controlling and eliminating circumstances judged to be a disvalue. The judgements are in no sense pragmatically neutral. Choosing to call a set of phenomena a disease involves a commitment to medical intervention, the assignment of the sick role, and the enlistment in action of health professionals (pp. 40-41).

Initially we believed Engelhardt's formulation to be admirably clear. But on second thought a nagging doubt arose. Innocuous words like "accounts" and "involves" do not easily catch the attention but, once they do, they tend to become elusive. Precisely what is an "account of disease"? A definition, a description, an explanation, an evaluation? Or all of this at the same time? How are commitments "involved"? In a logical way, in an empirical way, or in a normative way? Or in many different ways?

We do not know how Engelhardt would reply. If he would insist that the concepts of health and disease cannot be value-free in the sense that one cannot give value-free definitions (whereas other kinds of definition are possible) we would like to insist that he clarify his formulations. Just now we do not know where definitions enter the scene. Alternatively, Engelhardt could resort to some kind of holism. That could free him to some extent from the obligation to explain concepts and definitions in greater detail. But it would call for a defense of holism.

Suppose for a moment that holism of meaning is an ally of normativism. Where would that leave us? *Naturalism* is less likely to favour this kind of holism, since it accuses normativism of obliterating distinctions. So it is possible that the two positions are associated with different theories of *meaning*. For an adequate appraisal of the controversy, therefore, one will need some general ideas about meaning.

Holism has many varieties; we mentioned but two of them. It plays an important role in the philosophy of medicine (see e.g. Kopelman and Moskop, 1981, and Brody, 1985b). It is disregarded in the rest of this chapter. We will return to the subject in chapter VI, which will also deal with the allied *biopsychosocial* concepts of health and disease.

Whatever attitude one takes towards holism of meaning, *some* distinction between matters of logic and other matters will have to be made. Even in ordinary discourse such a distinction is continually made. This should be obvious from the difference between two common questions. One, what do you mean? Two, why do you think so? Normativists are often unclear about which kind of question they are answering. Engelhardt's writings (see various quotations in this chapter) are an example.

Some additional examples will strengthen our case. The philosopher Goosens (1980) has cogently argued that Engelhardt's version of normativism is too strong. For example, there can be cases of disease with nothing disvaluable (cf. Boorse's attack on strong normativism). Goosens himself defends a weaker form of normativism. In a crucial passage he argues as follows.

> Together, the nature of medicine and the nature of language make normativism extremely attractive, though not necessarily right, as a program for the analysis of medicine. Why, then, do particular normativist theses fare so poorly? One reason is that the right normative concepts have not been deployed, but the main reason is that there has been a too limited conception of how medical concepts could connect to normative ones. Some diseases, like arthritis, harm by their very presence. Others, like arteriosclerosis and the inability to make some kinds of antibodies, may produce no actual harm, but raise the probability of being harmed. Yet others, like tumors, may be dangerous for what they can lead to. What connects these medically significant conditions to well-being, however, is that they are *some threat* to well-being. ... Being a threat to well-being is proposed broadly as a necessary condition for negative medical concepts [not merely *disease!*], but not as a sufficient condition for any particular one (p. 107).

We agree with Goosens' attack on strong normativism as it is commonly defended in the literature. But his weaker variant is not well worked-out. Crucial to note is that the connection between medically significant conditions and well-being (cf. "What connects these ... conditions ...") need not be a *conceptual* one. Why not construe it as a factual relation? Ambiguity of seemingly innocuous words like "connection" makes Goosens' defense of normativism ineffectual. His attack on "neutralism" (naturalism) does not survive scrutiny either.

To show how nonnormative approaches go astray, let us take on [a particular version of] ... neutralism. The most widely endorsed approach to the analysis of health is based on the concept of functioning. Certainly knowledge of functioning is important to medical practice, simply because most negative medical conditions threaten their victims through disruption of functions. But, because there is no conceptual connection between proper functioning and benefit and harm, this approach is radically wrong for foundational analysis (pp. 110-111).

This argument simply begs the question. Goosens *presupposes* what he sets out to prove, that there must be a *conceptual* connection between health and benefit or harm.

Agich (1983) begs the issue in the same way.

On Boorse's view 'disease' ... is a description of a deficiency in typical species functions where 'function' means 'a standard causal contribution to a goal actually pursued by the organism'; my suggestion is that if the phrase 'goals actually pursued by the organism' is understood in social terms and in terms of freedom rather than biologically (since medicine concerns *human* disease), then the breadth of possibilities regarding disease as well as the value-laden character of disease judgements will become apparent (p. 37).

The argument shows that *if* one interprets "disease" in value-laden terms, one will get a value-laden concept. That indeed should not come as a surprise.

Similar problems arise, in a different context, when mental illness is discussed in a normativist way. Consider the following passage from Talcott Parsons (1981), whose analyses of the sick role are well-known.

Since it is at the level of role structure that the principal direct interpenetration of social systems and personalities comes to focus, it is as an incapacity to meet the expectations of social role that mental illness becomes a problem in social relationships and that the criteria of its presence or absence should be formulated (p. 58).

Are we dealing here with *definitional* criteria? Or is Parsons saying that mental illness is a social problem *as a matter of fact*? His article allows either interpretation. On the same page he writes: "Mental illness, then, including its therapies, is a kind of "second line of defense" of the social system vis-a-vis the problems of the "control" of the behavior of its members". What is the meaning of "is" in this statement? Remember that the verb "to be" is tricky! It may express a factual connection. The social system happens to react to mental illness in a certain way. Alternatively,

Parsons may want to say that certain reactions of the social system *constitute* mental illness.

Parsons writes as follows about the sick role.

Illness, then, is also a socially imputed generalized disturbance of the capacity of the individual for normally expected task or role-performance ... . Under this general heading of the recognition of a state of disturbance of capacity, there are then the following four more specific features of the *role* of the sick person: 1) ... he cannot be "held responsible" for the incapacity ... . 2) Incapacity defined as illness is interpreted as a legitimate basis for the *exemption* of the sick individual ... from his normal role and task obligations. 3) ... the sick person has an obligation to try to "get well" and to cooperate with others to this end. 4) ... the sick person and those with responsibility for his welfare ... have an obligation to *seek competent help* (pp. 69-70).

Again we are plagued by the ambiguity of "is", in the first sentence of this passage. As a consequence, the term "features" also gives troubles. Does it stand for criteria that define "illness"? Or does it refer to social norms concerning illness which, in point of fact, are now prevalent in our culture?

Normativists have rightly connected facts and values. But the nature of the connections needs much clarification.

4.2. *Making the best of science*

In the foregoing analysis, the controversy over concepts of health and disease was characterized in terms of logic and elementary philosophy. We now want to approach the matter from a different angle by concentrating on the role of science. This will be done by the presentation of various themes which illustrate how much research remains to be done. Some of them were already introduced in sections 2 and 3, but we will also present new material.

Science is more than biology. Boorse wants to have a scientific theory of health and disease. He simply *assumes* that it will have to be a biological one. Normativists seem to argue that such a scientific view of health and disease is not very helpful for medicine. Human values must be taken into account, so we cannot rest content with "pure" science. Normativists are actually making *various* choices. They introduce values *and*, at least implicitly, they give a more prominent role to psychology and social science. Part of the discussion, therefore, reflects a disagreement about the branches of science deemed relevant for medicine. Medical theory traditionally draws on biology. Should that be changed? Naturalists and normativists mostly do not defend their answer to *this* question. Chapter VI will deal with the issue.

Values play various roles. This point is connected with the previous one. Psychology and social science can deal with values "in an empirical way". They may shed light on normative aspects of health and disease without indulging in a normative stance (cf. our comments on the views of normativists). Studying values is not the same as endorsing them! Thus it may be possible to reconcile naturalism and normativism (see also section 3.2). Naturalists want science. Normativists want values. Why not have it both ways? We suspect that the primary issue between naturalists and normativists concerns the scope of *empirical* science in medicine rather than values *per se*. Naturalists tend to overconcentrate on biology. They seem to be afraid that the concept of disease will be overstretched if it is made to cover psychological and social problems. If the scope of medical theory is defined less restrictively, the study of values will have a legitimate place in medicine.

Medical nosologies are not that special. Biologists often quarrel about the purposes of classification (see e.g. textbooks mentioned in section 2.2; many papers in *Systematic Zoology*). In classifying organisms, one may want to establish a good basis for identification, or for the development of biological theories. If one concentrates on theories, one will have to make choices. Is evolutionary theory all-important? If it is, should classifications reflect evolutionary relationships between organisms? Classifications are obviously related to purposes. Perhaps one must allow for different kinds of classification suiting different purposes (see e.g. Kitcher, 1984).

Medical professionals often quarrel about the purposes of nosologies. In classifying diseases, one may want to establish a good basis for therapies. Or for the development of medical theories. If one concentrates on theories, one will have to make choices. Is epidemiology all-important? If that is so, should classifications reflect causal relationships between pathogens and hosts? Classifications are obviously related to purposes. Perhaps one must allow for different kinds of classification suiting different purposes.

The moral should be obvious. Discussions about values in medicine tend to suggest that medicine is special, that it cannot be like natural science (cf. Engelhardt's views and our comments on them). We would not agree. Problems which plague medicine are common in natural science as well.

Concepts of disease and concepts for diseases are very different. An analogy from biology may clarify the role of concepts for health and disease. In biological classification, the distinction of *categories* and *taxa* plays a crucial role. "Species", "genus", "family", and so on, are categories. Carrion crows constitute a taxon at the species level (a taxon belonging to the species category). Notice that some basic terms are ambiguous. The statement that there are many species refers to species in the sense of taxa! Not a few biologists and philosophers of biology have pointed out that such ambiguities easily lead to fallacious reasoning. By and large, however, biologists carefully distinguish taxa and categories.

The concept of disease is ambiguous in precisely the same way. So far, we have freely used it in different senses. But we were in good company. Both naturalists and normativists seldom discuss the distinction of "disease" (the category) and "diseases" (the "taxa"). On this score, they could learn a lot from biologists. The following example shows what we have in mind. One may rightly assume that rigorous, non-arbitrary definitions for diseases (the "taxa") are imposssible. The same goes for species. Does it follow that "disease" is an arbitrary category? By no means. Biologists generally agree that the species category is non-arbitrary! Perhaps the concept of disease behaves in a different way, but one should not assume that without argument. Close inspection of arguments in discussions about health and disease reveals that they are often incomplete. The arbitrariness of nosologies does *not* imply that "disease" is an arbitrary category. Anyway, the conceptual relations between "disease" and "diseases" require clarification (cf. Canguilhem, 1966).

Sophisticated concepts without theories are useless. In empirical science, discussions about concepts are mostly associated with comments on theories (and/or hypotheses, laws, models, etc.). The same is true, *mutatis mutandis*, of normative disciplines. How could it be otherwise? Concepts are meant to function in the context of theories, and understanding them will require some knowledge of the context. In our opinion, concepts are over-emphasized in Anglo-Saxon discussions about "health" and "disease". We would like to know characteristics of the theories associated with the concepts, and with concepts for diseases (cf. the distinction we made above). Otherwise attempts to develop a generally valid disease language will be rather pointless. Various authors, e.g. Macklin (1981) and Lazare (1981), have argued that medicine does not have a well-integrated body of theory; precisely this makes central concepts for health and disease problematic. Macklin thinks that such a theory can be developed in principle. With respect to diseases (the "taxa") we are sceptical in view of lessons learned from biology (other reasons for scepticism will be given in

Chapters V and VI; see also Brown, 1985a). An example will show what we have in mind. The logical features of species names may deny them a role in laws of nature and other general statements of biology (see Van der Steen and Voorzanger, 1986, and references in the article; also Ghiselin, 1987, and Mayr, 1987, and responses following their articles). Perhaps the study of particular species can only lead to "natural history", not to theories in a strict sense of the term. This is a real issue in current biology. How do matters stand in medicine? Would it be possible to elaborate universal laws and theories concerning diseases? Or must one be content with an analogy of natural history in biology? Gorovitz and MacIntyre (1976) have argued that medical science is indeed a *science of particulars* without universal laws. Schaffner (1986) has recently defended a similar thesis. But it does not follow that medical science is special. Biology, according to Schaffner, has similar features.

Current disagreements in the dispute over naturalism *versus* normativism can be resolved only if the role of science in medicine is elucidated. So one needs a characterization of "science" and "scientific theories". Views of science which are presupposed by the disputants may well be wrong.

Explications of concepts of health and disease (and concepts for specific diseases) need to deal with relativism. Boorse's views (cf. section 2) suggest that concepts for health and disease can have a place in straightforward biology. However, his biology is anything but realistic. Thus his typological approach made him minimize the impact of the environment. Once one gives the environment the role it deserves, concepts for health and disease turn out to be context-dependent even if they are confined to the domain of biology. For example, reference values used to characterize diseases make sense only if they are related to some environment.

Normativism can but increase the impact of context-dependency. To some extent, biological features of diseases may be invariant across environments. But norms and values stressed by normativists will show less invariance over (cultural) environments. So normativists will have to deal with cultural relativism. Most of them have not done that in an explicit way (notable exceptions are Whitbeck, 1981, and Brown, 1985a). Research on relations between culture and disease has shown that relativism is a profound issue. Specifically, it is now hardly possible to give a trans-culturally valid characterization of psychiatric disorders (see e.g. Kleinman and Good, 1985).

The survey of problems presented above suffices to show that there are no reasons now for accepting either naturalism or normativism. Current disputes are too simplistic. They must be replaced by a new research program.

5. THE INTERPLAY OF SCIENCE, COMMON SENSE AND PHILOSOPHY

The upshot of the analysis in the foregoing section is that current discussions between naturalists and normativists are off the mark. They would seem to call for a new research program in philosophy of medicine. The program we have outlined is ambitious, and so it should be *if* one wants to solve the riddles uncovered. But it is a limited one since the riddles themselves represent a one-sided way of looking at medicine. We have concentrated so far on the impact of science and philosophy of science on medicine. Medical practice got little attention. The present section deals with relations between medical theory (science) and medical practice, still mostly from an Anglo-Saxon perspective.

As matters now stand, the search for *one* general concept of health and *one* general concept of disease is futile. The reason is simple. Unambiguous concepts (sensible ones, that is) can only be articulated in the context of a coherent theory, and medicine does not have such a theory (see section 4.2). Neither has the philosophy of medicine. To make things worse, concepts of health and disease play rather different roles in medical theory and medical practice, respectively, and the same goes for nosologies. This in fact is what makes discussions between naturalists and normativists so untractable. Naturalists concentrate on theory, normativists on practice. No wonder that they end up with different concepts.

Precisely how can *practice* be involved in the conceptualization of health and disease? There is a tension here because one needs *theory* for *articulated* concepts. To answer the question one should come to grips with relations between medical science (theory) and medical practice, and that is by no means easy. To some extent at least, medical practice is applied medical science. Now *current* medical science is mostly biology. Should we remedy its one-sidedness by paying more attention to medical psychology and medical sociology *besides* biomedicine (see e.g. Mechanic, 1986)? Or must current theories of medicine be replaced by integrative theories (cf.

psychosomatic medicine, which is discussed in chapter VI)? Perhaps medical practice does need more science, or better science. But it also needs more *than* science. Would common sense suffice as a supplement (cf. the excellent textbook by Bloch, 1985, who deliberately introduces "common sense psychology" in medical physiology)? Or do we need philosophical theories since science cannot adequately deal with *persons* (cf. anthropological medicine, as discussed in chapter II)? Such issues are all connected with "the" mind-body problem, which has haunted western science and philosophy throughout the centuries. The problem will be discussed in chapters V and VI. In the present section we will concentrate on the more general issue of relations between science, common sense and philosophy in the context of medicine.

The relations between science, common sense, and philosophy are intriguing. Various traditions in philosophy are intimately linked up with common sense. Phenomenology resorts to primary experience to redress the one-sidedness wrought by the abstractions of science. In the Anglo-Saxon tradition, *ordinary* language philosophy tries to solve perennial problems by explication of normal discourse. Unfortunately, research in psychology has shown in the last few decades that common sense is extremely unreliable (see chapter III, section 5.3; chapter V, section 4). "Plain facts" which we come across in daily life are never really plain. They are coloured by notions of *folk science* (and fragments of popularized science). Such notions are often problematic. So science will have to correct common sense. But scientists are ordinary people; in their work they will have to rely on common sense as well. Faulty reasoning, biased interpretation and ambiguity are common in science. Fortunately philosophy (especially methodology) has developed techniques to redress shortcomings of science. And so forth. The point will be obvious by now.

Explications of concepts given by philosophers of medicine are often in the style of ordinary language philosophy; the book by Culver and Gert (1982) is a typical example. Normativists in particular have adopted this approach. They often arrive at concepts of health and disease that represent ordinary discourse. That should explain the ensuing lack of consensus, or partly so. In order to be useful, ordinary discourse needs some vagueness and ambiguity, so it easily becomes a source of disagreements. One can of course try to improve on common parlance (and folk science associated with it), but then one will need coherence with some theory: concepts outside a theoretical context can hardly be useful. Biology will not suffice for the normativist, academic psychology often fails to address practical issues, and ordinary language philosophy is *ipso facto* tied to common sense. It should indeed be very difficult for normativism to get rid of vagueness and ambiguity.

Naturalists are less ambitious. They wisely stay within the domain of biomedicine (where they encouter other problems). Within that domain they will not come across the problems which plague normativists. But they will have to face them when they try to make the connection with medical practice (see the remarks on values in section 4.2).

There is now much disharmony between medical theory and medical practice. We will give various examples concerning the concepts of health and disease, and nosologies, to illustrate this. For a change, we will concentrate on views expressed by representatives of the medical profession rather than those of philosophers.

Waddell et al. (1984) critically analyse the treatment of patients with chronic backache by orthopedic surgery. They notice that treatment of chronic pain is often determined more by the patient's distress and demand for help than by the severity of any physically demonstrable disease. Thus hypothetical disorders lead to progressively more dangerous and damaging treatments without positive results. The study of Waddell is aimed at improving practice. The patient must be treated, not the disease. So one needs a correct interpretation of signs and symptoms.

The authors proceeded as follows. A large group of patients was studied with respect to treatment previously received, objective symptoms and signs of physical disease, and inappropriate illness behaviour. "Inappropriate illness behaviour could ... be recognized clinically as illness behaviour out of proportion to the underlying physical disease and related more to associated psychological disturbances than to the actual physical disease" (p. 739). Inappropriate pain is an example. Objective assessment of the severity of the physical problem was based on characteristics such as lumbar flexion, straight leg raising, signs of root compression, etc. Statistical analysis of the results showed that patients showing a large amount of inappropriate illness behaviour had indeed received significantly more treatment. The moral is clear.

If ... a patient's distress and illness behaviour are not recognised they may be misinterpreted in terms of a hypothetical and unconfirmed physical diagnosis. If apparent severity and failure to respond to simpler treatments are then used to justify invasive or dangerous treatment such as surgery, that misdirected treatment is not only foredoomed to failure but may actually reinforce and aggravate the illness behaviour. The conclusion is simple: the symptoms and signs of illness behaviour must be distinguished from those of physical disease (pp. 740-741).

The authors' approach is fully justified *with respect to* their aim to prevent excessive invasive or dangerous treatments. But notice what it presupposes. Theories and data of orthodox medicine are used as overriding criteria in the interpretation of complaints. (Moreover, *negative* evidence is given a crucial role.) "Inappropriate" complaints thus are complaints that do not fit orthodox views which are taken for granted. One is left without guidelines to deal with such complaints.

Higgins (1984) deals with similar problems in a more subtle way. He is also concerned with patients who have symptoms suggesting physical disease but in whom no physical disease can be found; his analysis is not restricted to a particular area of medicine. Especially individuals without evidence of psychological disorder are his concern. Higgins points to inadequacies of conventional medicine. He argues that medicine needs to concentrate on cognitive aspects of symptoms. There are no uninterpreted symptoms. Everyday perceptions, experiences and understanding which reach consciousness are always structured by cognitive schemata. So it should not come as a surprise that models used by clinicians often diverge from recognized theories.

A cognitive theory of health and illness may foster a better understanding of patients. But it can do much more than that. One should also consider "the cognitive structures given to doctors by medicine and medical training, and modified by experience in practice" (p. 736). True, doctors know well-defined clinical entities which merit the term "disease". But medicine must also deal with phenomena in the range between "real" diseases and non-disease. "Faced with all this it is impossible for us to define disease (as we currently use this term) as something having an existence independent of interpretive activity" (ibid.).

"Medicine and society" is another area which can benefit from a cognitive approach.

... we have to realize that clinical practice has much in common with the cognitive work done by lay people and that medical knowledge is simply a 'selective schema whose elements owe their place to their pragmatic utility' (Bloor 1976). A basic assumption of that schema is that symptoms are useful indicators of biological events. That assumption drives us to contortions of speculative thinking in order to explain how social factors and stress lead us to illness. An alternative assumption is to regard symptoms as indicators not of a medical problem but of a disorder that may be wholly social or wholly medical or something in between. This would lead us to concentrate on the mechanisms which lead the individual or the doctor to interpret them in terms of illness. We may then begin to understand why so much in medicine is not what it seems (p. 736).

Higgins' approach is hardly compatible with that of Waddell et al. The latter authors are in fact also concerned with cognitive interpretation, but they seem to presuppose that the core of medicine (theories, data) does not need it. We think that they trust "objective" orthodox medicine overmuch. Higgins' takes a relativistic position which leaves much more room for common sense; when he uses the expression "cognitive *theory* of health and illness" he is apparently not thinking of plain science. We have already defended a similar position in commenting on alternative medicine (see chapter III, especially section 5.3).

Many other authors in the field of medicine have recently been concerned with complaints of patients which do not fit nosologies. For example, Jennings (1986) argues as follows. Much clinical confusion might be clarified if separate terms are used for bodily pathologic change (disease) and experienced suffering (illness). (Jennings attributes the distinction to Barondess, 1979; in fact it had been made long before by *philosophers* of medicine.) Illnesses are in two categories. Some arise from disease (medical illness), others from personal difficulties in living (existential illness). The distinction is meant as a pragmatic one, it is not based on Cartesian dualism. There are many examples of existential illnesses which are wrongly treated as diseases. Chronic brucellosis, obesity and hysteria are examples. Their being treated as diseases results in medicalization.

I suggest that it would serve both physicians and patients to maintain conceptual clarity when personal suffering is being investigated. Suffering arising from bodily disease can then be properly dealt with by physicians trained in the methods of biomedicine. Other suffering can be properly treated by anyone, physician or not, who is sufficiently skilled (p. 869).

We think that this oversimplifies a bit, although we like the emphasis on common sense. The dichotomy which Jennings introduces can hardly be as neat as he suggests. Cartesian dualism does not cease to be a problem if one transforms it into a pragmatic variant!

McWhinney (1986) thinks that we need a new clinical method which concentrates on empathy and attentiveness, in order to elaborate better connections between pathology and the signs and symptoms uncovered by clinical medicine. Illness can only be understood if one gets to know the patient's inner experiences, and the classical question-answer method of interviewing (which is doctor-centred rather than patient-centred) gives biased results. McWhinney mentions research by Donovan, a gynaecologist with training in psychiatry, who used an open-ended, patient-centred method. When he used this method with patients believed to have menopausal problems, he found no evidence of menopausal causation; the symptoms were an aspect of personal problems, often with a long history. Donovan's work is often cited in behavioural science, but it had no impact on gynaecology. Other attempts to reform clinical method also remained ineffective. McWhinney argues that this is explained by the lack of validation criteria for patient-centredness.

All these authors are concerned with discrepancies between theory and practice, but the strategies they propose for dissolving them are dissimilar. Is current practice to be blamed, or are biomedical theories and nosologies inadequate? Waddell et al. rather dogmatically seem to blame practice. We cannot accept the style of their approach; additional comments are given below. Higgins lets criticism go both ways. We agree with his thesis that a cognitive approach of health and disease is useful. And we would add that many other approaches aiming at a relativistic perspective on medicine may be equally profitable (cf. medical psychology in a wider sense of the term, anthropological approaches of medicine and culture, general methodology). It will indeed not be easy to make the best of the plethora of approaches now proposed. Jennings and McWhinney, each in his own way, aim at improving practice. But they seem to suggest that nosologies should not remain unaffected.

By and large, one is left with the impression that theories and nosologies of biomedicine are less easily changed then practice. If there are discrepancies (cf. complaints not fitting nosologies), one can remove recalcitrant data from the domain of biomedicine (Waddell et al.'s inappropriate complaints; Jennings' existential illness), or one can improve on the data by a different approach of patients (Higgins' cognitive approach; McWhinney's patient-centred approach). All this suggests that it is easy to shield biomedical theory from adverse evidence. Would it be possible to use data from medical practice as a *test* of biomedical theories and nosologies or as a basis to change them? We will approach this question, in a general way, from a philosophy of science angle.

A discrepancy between a theory (and/or nosologies associated with it) and signs and symptoms of patients could point to various kinds of inadequacy of the theory. Firstly, the theory could be inadequate in the sense that it does not suit particular purposes. Secondly, the theory could be false. We will concentrate on the second kind of inadequacy, the stronger one. Consider again the approach of Waddell et al. A theory (actually various theories) and "facts" ascertained by physical examination of a patient jointly lead to the prediction that she should not have pain. But she does have pain. Should that militate against the theory? The answer given by Waddell et al. is emphatic. The patient's pain is "inappropriate". That seems a dogmatic line to take. But of course there is no warrant either for the conclusion that the theory is in trouble. Logically, the situation is rather complicated. For one thing, the prediction could only be inferred from theory and facts by the covert use of auxiliary theories. Biomedicine in a strict sense of the term can say nothing about pain *experienced*. So Waddell et al. must have used an implicit theory concerning relations between "the mental" and "the physical". The chances are that it is an implicit, unarticulated theory.

Which theory should one blame in the face of adverse evidence? There is no easy answer. One could argue that unarticulated theories had better be rejected if there has to be a choice. Unfortunately, this would sever the connecting link between (biomedical) theory and practice. Decisions on the value of "evidence" from practice would so become impossible. Waddell et al. understandably do not accept that. One has to keep one's work going. They apparently accept some implicit auxiliary theory.

There must be a second auxiliary theory as well, otherwise the pain could not be called "inappropriate". The authors obviously assume that pain can also be caused by factors outside the scope of their own theory. These factors, which are left unspecified, get the blame if predictions go wrong (an *ad hoc* move, unless there is independent evidence pointing to effects of such factors). At a more general level, there is the additional presupposition that the two classes of factors which can cause pain are to some extent independent, otherwise the strategy followed by Waddell et al. should collapse. So again some mind-body theory will have to be involved.

The situation we described is the normal one in research, pure or applied. We have simply made explicit what is mostly left unsaid. But in this particular case there are good reasons to be explicit precisely because orthodox medicine is nowadays heavily criticized. And arguments concerning its merits and demerits are likely to be sterile unless one is aware of intricacies (see also chapter III). Research in biomedicine and treatments in medical practice must of course continue. That is possible only if one ignores many problems (preferably not the same ones continually). Ignoring them may be healthy as long as one is willing to admit ignorance.

Would it be possible to use data from medical practice as a test of biomedical theories and nosologies? That was the question we started with. The answer is that practice can be brought to bear on theories, but not in the sense that any theory is shown to be true or false. Theories are never testable in *that* way (see chapter III). Testing is a matter of prudentially weighing evidence. One will have to take lots of assumptions for granted. As evidence for or against a theory accumulates rational decisions concerning its value will at times be possible, especially if various assumptions have been tested in an independent way. But more often it will be wise to reserve judgement.

In medicine, the connections between theory and practice are often weak; the auxiliary theories one needs are seldom even articulated. *Here we would like to locate an important task for the philosophy of medicine.* Philosophers should be able to provide better *reconstructions* of biomedical theories, and of links between theory and practice. Full-fledged reconstructions are seldom presented in the literature, and they are badly needed. An appraisal of a theory is possible only if it is sufficiently clear in the first place.

Adequacy of theories, as we have already noted, is not a matter of truth or falsity alone. A theory can be totally inadequate with respect to the purposes one would like it to serve even though there are lots of reasons for regarding it as true, or at least as *empirically* adequate with respect to some domain. This kind of inadequacy is very important in relations between medical theory and medical practice. How to deal with it? As far as philosophy of medicine is concerned, we would again like to give reconstruction a prominent place on the agenda. Sensible policies are possible only if one has a clear picture of theories and their limitations, of connecting links with practice, and of the pragmatic context.

The most important benefit of philosophical reconstruction, perhaps, is that it may reveal the vastness of our ignorance. It has been argued that unwillingness to admit ignorance may lead to dangerous forms of medicalization. The treatment of asthma in children is an example. Investigations by Roe (1984) suggest that "with our obsessive application of the scientific method and neglect of the art of medicine it has been transformed from a relatively benign disease into one with a considerable mortality" (p. 389). The prescription of potent drugs is presumably one of the sources of increased mortality.

Almost half a century has elapsed since Pfaundler and Schlossmann asserted, "However distressing may be the single attack it is transient even if not treated" ... . This study supports this view that the natural history is indeed benign.

How is it, then, there is almost universal belief the condition requires aggressive therapy with dangerous drugs? How is it that this belief persists despite the fact that its implementation has not only failed to reduce the mortality rate but has actually been accompanied by a rise in the rates in certain countries and the emergence of a new unexplained mode of death? How is it more credence is given to the hypothesis that deaths are more commonly due to undertreatment than to the equally valid hypothesis that they are the result of overtreatment? ...

Such an absurd and dangerous state of affairs is surely a manifestation of a very deep-rooted bias. It is submitted that this bias stems primarily from man's reluctance to admit to ignorance, from his preference for acting on unsubstantiated hypotheses rather than doing so - his "obsessive mythopoesis" and secondarily from his faith in a false philosophy: a philosophy which regards man as no more than a machine, as something that can be understood in mechanistic terms (pp. 396-397).

Unless there is awareness of ignorance, medicine will not be able to excercise restraint and modesty, two virtues which it badly needs.

So far we have hardly paid attention to mental health and mental disorders, and psychiatry as a source of theories for dealing with health problems. In chapter VI we will deal in a general way with "the mental" in medicine. Here we will only give a few comments on special problems associated with concepts of health and disease and nosologies. There are indeed special problems for lack of generally accepted *theories* concerning "non-organic" disorders.

Many authors (e.g. Muscari, 1981; McGuire, 1986; Schwartz and Wiggins, 1986a; Stephens, 1986) have argued that DSM-III, the current Diagnostic and Statistical Manual of Mental Disorders of the American Psychiatric Association, is unsatisfactory since it is not associated with scientific theory. The emphasis is on a descriptive clustering of signs and symptoms which is *operational*. DSM-III is obviously a reaction to earlier attempts to give nosologies a theoretical basis. Such attempts never led to consensus as they mixed science with philosophy.

As Schwartz and Wiggins have argued, the emphasis on operationism was inherited from logical empiricism, which dominated Anglo-Saxon philosophy of science a few decades ago. One of its representatives, Hempel (1965) has strongly advocated the use of operational concepts in psychiatry, and various psychiatrists (notably Kendell, 1975, 1983, 1984) have followed his lead. However, Kendell has significantly modified Hempel's views (see Schwartz and Wiggins, 1986, p. 112). Hempel argued that operational definitions must be linked with (empirically relevant) *theory*, otherwise they have no solid base. In the absence of adequate theories one must try to develop theories and operational concepts jointly. Thus one will need accounts of etiology to make sense of nosologies. Kendell, however, dispenses with etiology. He focuses exclusively on symptoms since etiology is poorly understood.

Schwartz and Wiggins' point is that once the influence of operationism on current psychiatry is uncovered, the feasibility of alternative appproaches is more easily recognized. They would like to put more emphasis on intuition and empathy which are apparently depreciated by psychiatrists like Kendell (their own sympathy is with philosophical traditions of continental Europe; see Schwartz and Wiggins, 1986b). We have some additional comments. Recent Anglo-Saxon philosophy of science (see the survey in section III.3) is hardly compatible with the spirit of operationism. The links between theories and the level of observations are indirect and complex, they cannot really take the form of operational definitions. But concepts for mental disorders will have to be connected with theory in *some* way. Unanchored, freely floating nosologies like those of DSM-III will remain adrift as there are no guidelines for judging the importance of symptoms.

It is remarkable that ideas from the philosophy of science have so much influence in science long after philosophers have ceased endorsing them (see also chapter II, section 2.2, and Van der Steen, 1982). Operationism is by no means limited to "taxonomy" in psychiatry. Taxonomy in biology also has its share. Hull (1968) has already warned against operationism in biology some two decades ago!

Nosologies are one source of problems. At another level psychiatry must deal with (mental) health, disease and illness as general categories. Needless to say, the categories are problematic. There is no generally accepted theory as an organizing framework for central concepts. This presumably explains the impact of ordinary language philosophy in attempts to clarify the notions of health and disease in psychiatry. Thus Brown (1985b) has tried to give explications on the basis of Culver and Gert's (1982) ordinary language approach. He could but admit defeat. The most promising explication of "mental illness" in terms of disability ultimately shows that "control (or treatment or excusing) of deviance is not justified because people are disabled; rather, disability is ascribed to people to justify the control we feel obliged to exert" (p. 575).

What theories do we need in psychiatry to arrive at adequate concepts and nosologies? One answer is that such theories will have to deal with *persons*. So biology will not suffice. Psychology and/or philosophy are indispensable (Muscari, 1981). But biological psychiatry has its advocates. An evaluation of various standpoints will be given in chapter VI.

6. AFTERTHOUGHTS

Our philosophical "style" in this chapter mostly had the flavour of Anglo-Saxon philosophy of science. This should not be taken to indicate philosophical preferences. We do have sympathy for other philosophical traditions such as phenomenology. So why the one-sidedness? The reason is simple. Analysis of concepts typically is an Anglo-Saxon affair.

Anglo-Saxon discussions about the concepts of health and disease, and the status of nosologies, of course belong to a cultural setting which needs to be put in perspective. Disagreements between naturalism and normativism over values ultimately reflect commitments concerning the place of science in medicine, and the role of medicine in society.

We have uncovered shortcomings of naturalism and normativism by an analysis of arguments rather than basic commitments. Analysis cannot handle commitments. But it can bring them to the surface so that one better knows the stakes involved.

Our impression is that the naturalist and the normativist are both fighting a battle against the medicalization of western society. They concentrate on different aspects of medicalization, and that explains the differences in their philosophies. Normativists argue that the whole edifice of medicine is permeated with values, as it should be. Naturalists do not accept that because it tends to make medicine a moral enterprise. If health and disease are normatively defined, medicine easily becomes a cultural force without clear boundaries. Thus medicine directed at the promotion of health in the sense of inclusive well-being is easily invested with responsibility beyond competence.

Naturalists therefore want to restrict the domain of medicine. Medical theory must not be tainted with values, it must must take the form of objective science, and practice associated with it must remain *medical* practice. Normativists do not accept that because it tends to make medicine a dehumanizing enterprise. If health and disease are defined in terms of science, medicine easily becomes a force without cultural boundaries. Science and allied technology have a dominant impact on our perceptions of health and well-being. As medicine gives them a central role in the promotion of health, it easily appropriates responsibility beyond competence.

Which is the right view? The answer is, neither. Or both. Naturalism has got a point. So has normativism. But the battle they fight against each other is futile. They had better join forces to fight medicalization, and foster modesty in medicine.

Ours is a culture without a cohesive world-view. As science has taken a materialistic turn religions have lost foothold, or have been put in a separate realm. No unifying force is left to satisfy the spiritual needs of man. Thus the vacant seat of religion is easily occupied by substitutes that cannot play its role. That, perhaps, is why medicalization is so pervasive.

Let us be realistic about medicine. It does contain empirical theories, so it deals with facts. But it is also associated with values. Medicine does not have one integrative scientific theory for the domain of facts. There are lots of loosely interconnected theories, some of them apparently well-confirmed, others problematic. And of course there are many gaps in our knowledge. Discursive knowledge without gaps is impossible. For the realm of values, there is no integrative theory either.

It would not be easy to reverse the trend of medicalization because it is deeply embedded in the structure of western society. Could philosophy help? In principle it could in two different ways, one ambitious, the other modest. The ambitious way is to develop a world-view which puts medicine in the proper place. The modest way is to uncover limitations of medicine by painstaking analysis. We feel unable to develop any ambitious program, so we have concentrated on the second alternative.

CHAPTER V. MIND AND BODY IN SCIENCE AND PHILOSOPHY

1. INTRODUCTION

In western medicine, approaches to mind-body relations have changed considerably during the last two centuries. In the early 19th century, sensations, emotions and bodily phenomena were all still covered by physiology. Only after the 1850's, under the growing impact of the natural sciences, did physiology get its modern restricted scope. Medicine, in the same period, almost developed into a branch of biology which tends to disregard mental and social aspects of human life.

During the last 60 years, however, the visible limitations of this approach forced medicine to accomodate findings of psychology and, later, sociology. The study of the vegetative nervous system and of the endocrine organs in the first decades of this century, seemed to bring large parts of psychology under the dominion of the natural sciences. Almost simultaneously, psychoanalysis developed a model of mental functions with biological extensions. In the 1930's, confluents of these mainstreams merged into psychosomatic medicine. This movement started out on an ambitious program for integrative explanations of human diseases. However, the accelerating progress of biochemical and pathophysiological research since the 1950's has tended to outstrip it. Psychosomatic medicine helped to establish medical psychology and sociology as legitimate components in education and research. And it still stimulates the study of interactions between the mental, the social and the somatic. Today, the multidisciplinary approach to matters of health and disease is common practice, but no unitary system of explanation has emerged.

Meanwhile, developments in medicine (and biology, psychology) have influenced the philosophical study of the mind-body problem. It would have been foolish to disregard the wealth of data generated by brain research and neuropathology, and psychiatry. Computer science also suggested new approaches to the mind-body in the last few decades. The problem is now addressed in many ways by different philosophers. In the

118

present chapter, we will examine a sample of divergent philosophical views on the problem. This will set the stage for a more specific evaluation of the mind-body problem in medicine, which is given in chapter VI.

After presenting some philosophical background, we will consider recent theories in the Anglo-Saxon tradition which aim at solving "the" mind-body problem by appealing to brain research. We will argue that biology does not suffice as a scientific basis for philosophical mind-body theories. Next, we will show that psychology, and integrative views which combine various fields of science, are not very helpful either. Approaches which developed in continental Europe, especially phenomenology, will be considered subsequently. The relevant authors, many of them physicians or psychologists, started from conscious life or human existence as their primary datum. Unlike most Anglo-Saxon philosophers, they regarded the mind-body problem as an artifice, necessary for the scientific study of man but obstructive to philosophical thought as well as moral action. We will argue that phenomenologists have surely got a point, though their philosophy has limitations of its own.

2. THE PHILOSOPHICAL AGENDA

2.1. The spectre of integration

The mind-body problem in the West is as old as philosophy. And all the "solutions" which have been proposed are still with us. There is a confusing array of labels for characterizing mind-body theories (see e.g. Bunge, 1980; Churchland, 1984; Wuketits, 1985). Just now we will only mention distinctions which play an important role in sections 3 and 4.

The primary distinction to be made is between dualism and monism. Both come in many varieties. Dualism may either take the form of *substance dualism* (mind and body are essentially different entities) or *property dualism* (there is no fundamental distinction between entities; an entity can have two kinds of property which are quite different). Other distinctions concerning dualism will be introduced in section 3.

Monism also takes various forms. First of all one should distinguish *materialism*, *neutral monism* and *spiritualism*. The labels, rough though they are, will not need explanation. Materialism is now the most popular option. Within psychology, *behaviourism* is the classical variant. Nowadays (the predominant materialistic variant of) *functionalism* has more adherents. It assumes that mental states are functional states of the brain.

Various reactions to the persistence of controversies concerning the mind-body problem are possible. First, resignation. Problems need not always be solvable. Second, rejection of the philosophy which apparently generated the problem. Perhaps philosophies of other cultures have a solution. Third, declare one's own solution the sole victor in a dogmatic way. Fourth, depreciate philosophy and turn to science or at least philosophy inspired by science.

The first reaction is uncommon. The second one has been adopted by many scholars, philosophers and scientists alike (cf. the New Age movement). The third one is still common in philosophy. The fourth one is perhaps the most popular, both in science and in recent philosophy which tries to develop theories with science-like qualities. In the present chapter we will defend the thesis that the fourth option is no good. Modern science seems to shed new light on the problem, but appearances are deceptive.

Why should people expect so much of science? That is easily understood. Progress in neurobiology is impressive, and neurobiology is concerned with mental processes like learning, memory, and what not. And we have a science of the mental, psychology, which is still so young that it merits the benefit of the doubt. New disciplines may well provide new answers. Moreover, there are signs that it will be possible to integrate these domains of science and it is thought that this will *ipse facto* lead to an integrative view of the mental and the physical.

The ideal of a unified science is obviously an important issue for those who would argue that science will give one the solution of the mind-body problem. It is a venerable ideal which has been a source of inspiration for logical positivism a few decades ago. Unity of science was then one of the most important themes in *philosophy*. It isn't anymore. How come? *Recent* philosophy suggests that the ideal in its original form cannot be implemented, it is not realistic. It was originally thought that integration in science will come about through *reduction*. More specifically, sciences concerned with higher levels of organization were thought to be reducible (in principle) to physics, the most basic discipline. Biology should be reducible to physics, *psychology to biology*, and so on. However, reduction turned out to be a problematic notion.

In spite of the demise of the original program for a unified science, reduction is still the most important paradigm of integration. For this reason we will discuss the fate of the unity of science ideal in greater detail.

The concept of reduction which played a crucial role in early logical positivism has been articulated in many publications. The *locus classicus* is Nagel (1961). What does is mean for a theory to be reduced to another one? The question can be answered only if the meaning of "theory" is sufficiently clear. Nagel adopts a conception of theory from the logical positivists which is now known as the received view. On this view, scientific theories consist of general statements, laws of nature, with deductive relations among them. The laws themselves are at a very abstract level, so they cannot directly express what happens at the level of observations. Special principles called correspondence rules are needed to connect the language of theories with the language of observations. On Nagel's view, some theory T is reduced to a theory T* when the laws of T are logically derived (deduced) from T*. This is possible only if the concepts of the two theories are connected by definitions or other links. T* may have the role of a (more comprehensive) successor of T. In another case, T* and T may be concerned with different levels of organization (T* with the lower one). The two cases need not exclude each other.

Many different attacks have been launched against the general reduction paradigm outlined above. For example, it has been argued that putative "reductions" will always lead to a revision of the theories involved. Then we will have replacement rather than reduction in the sense of logical derivation (see e.g. Kuhn, 1972; Feyerabend, 1975, 1978; their theses have been discussed in numerous books and papers). There have been attempts to salvage the reduction paradigm by making it more sophisticated. Schaffner (1967, 1976), for example, has developed a model which allows for a modification of theories in the process of reduction. All the new variants of the classical reduction paradigm, however, will have to face other challenges as well.

Reduction which involves different levels of organization will require neat links between levels. Now it is conceivable that classifications which are needed at different levels simply do not allow the elaboration of neat links. In technical terms, concepts denoting entities at different levels may be connected by many-many relations. This would make reduction problematic. Take "the mental" and "the physical", putatively at different levels of organization. Would there be a one-one relation between, say, types of brain state and types of thought (e.g. thinking kindly about someone)? That seems implausible. Different types of brain state could be associated with the same type of thought, and one type of brain state may be compatible with different types of thought. Thus we will have problems if we try to reduce psychological *theory* to biological theory.

Hull has argued in many publications (see Hull, 1981) that this is a situation which obtains even in genetics. At first sight, genetics is the discipline *par excellence* which should allow reduction in biology. But not so. It is true that a few decades of research have resulted in links between classical (non-molecular) genetics and molecular genetics. But if Hull is right, many-many relationships are pervasive, so that reduction is not really feasible.

It has been argued that properties at higher levels of organization may be identified with ("reduced" to) lower-level properties even when one is plagued by many-many relations. Perhaps a particular combination of higher-level properties will be present whenever there is a specific combination of lower-level properties. The number of properties involved could be very large, so that relationships between different levels will not be "neat". Reduction of laws will then be precluded. In technical terms, we will not get a type-type reduction (a higher-level property of a particular type will not correspond with one type of lower-level property), but a token-token reduction. Higher-level properties in this case are supervenient on lower-level properties. Various philosophers have tried to solve mind-body puzzles in terms of supervenience (e.g. Kim, 1978). The same goes for conceptual problems of evolutionary biology (Rosenberg, 1978, 1980, 1985; for critical comments see Van der Steen, 1986a).

Token-token reductions are very interesting *philosophically*. But *ex hypothesi* they will not lead to integrative *scientific theories*. The feasibility of such reductions, therefore, is rather elusive (see also Kincaid, 1986).

In short, orthodox views of reduction are problematic since the concept of "reduction" is rather untractable. But that is not all. The explication of "theory" elaborated by logical positivists (cf. the received view mentioned above) came under attack as well. There is now a serious competitor called the semantic view (Sneed, 1971; Stegmüller, 1973; Suppe, 1977), which dispenses with correspondence rules and other devices developed by received view theorists. On the semantic view, theories consist of "ideal systems" without empirical content. Applications of theories (i.e., the identification of "empirical systems" with "ideal systems") rather than the theories themselves say what the world is like. Various philosophers of biology (e.g. Beatty, 1980; Thompson, 1983; for additional references see Sloep and Van der Steen, 1987) have argued that this is the better view as far as biological theory is concerned. We are not convinced (see Sloep and Van der Steen, 1987, and the discussion following the article).

Reduction (in any sense of the term you may prefer) is but one possibility to integrate scientific theories (see e.g. Darden and Maull, 1977; Bechtel, 1986). However, most philosophers of science have over-concentrated on the development of models for reduction. A philosophical inventory of different kinds of integration which play a role in science is badly needed.

The philosophical scene we portrayed would not be very relevant *if* science itself would provide a paradigm of integration. Then one would have good reasons for a new defense of the unity of science ideal. However, science is not now manifesting much unity. The point is that there are *many integrations* of disciplines. We do not seem to be getting less disciplines, we are getting more of them. Now one could argue that there are *philosophical* reasons for expecting that an integrative science will develop in the future. But then we would be back where we started. Philosophy does not have a new paradigm.

True, there are *programs* for a new paradigm. General systems theory, as developed by Von Bertalanffy (see section 3.5) is one of them. However, this program has not really been implemented in science. It has remained a means to talk *about* science.

How will all this bear on medicine? Orthodox medicine contains much science in the form of *biology*. There are many who have argued that biology alone is not enough. One needs psychology as well. Medicine would obviously benefit from the development of integrative theories. Indeed there are attempts to develop such theories, e.g. in psychosomatic medicine. What would the status of such integrations be? Do they represent reductions? Or a new form of integration which philosophers have failed to describe? Or is this not really an example of integration? These are important questions. Much to our surprise, they do not get much attention in the philosophy of *medicine*. One has to turn to other areas of the philosophy of science to learn how one could approach such questions. We would suggest that philosophers of medicine should give them more attention. True, medicine is *more* than science, but it does contain science. So there is a reason to associate philosophy of medicine *to some extent* with general philosophy of science.

2.2. How to proceed?

Extreme specialization in science cannot be very fruitful if adjacent areas remain unconnected. One needs integration besides specialization. However, attempts to achieve "grand" integrations which cover the whole of science are misguided. They will backfire in most cases. One should not try to understand the behaviour of cats, the breakdown of cars, and love-affairs from a single perspective.

Good theories are needed, but why *grand* theories? If one is in search for the meaning of life, one should turn to religion, not science. If one is ill, the search for treatment may make one consult a doctor, a psychologist, or both. They will look at etiology in different ways, and they may prescribe very different treatments. Whose advise should one accept? There is no general answer. The relevance one ascribes to various etiological factors will depend on the acceptability of theories from biology and psychology, and of the methodology used in testing them. Newton's laws will not matter *in this context*. If one is not sure about a sensible course of action, it may be wise to *go shopping* in science without staying too much in one particular place. Science resembles a market-place rather than a cathedral.

Scientific theories have obvious limitations which are often forgotten. Scientists have a tendency to pass the boundaries of their own discipline without noticing that they are outside their territory. Biology seems to be in a very good position for transgressions. It deals with many levels of organization, and this in itself suggests that there are few things which are not covered by biology in some way. Moreover, it has the advantage of being a natural science without much foundation problems whereas it is still entitled to develop a view of man. It has been thought that (evolutionary) biology is a good basis for ethics, for epistemology, etc.

We do not think that biologists have been sufficiently cautious in overstepping the boundaries of their own field. Researchers in theoretical psychology and the philosophy of psychology have shown in the past few decades that we have no adequate theories of mental processes. They have explicitly discussed many philosophical problems which will have to be solved before it will be possible to elaborate any adequate theory. Now biologists have developed theories of their own without bothering about the literature produced elsewhere. They just assimilated some philosophy. It is fairly easy to put their work in the proper perspective once one knows what is happening here. It seems to us that psychologists would not be seduced as easily to trepass on foreign grounds. They have long since been acquainted with foundation problems which biologists seldom bother about.

The present chapter has two functions. Firstly, it will provide a survey showing how science and philosophy deal with the mind-body problem nowadays. Secondly, it will present a critical analysis of views inspired by science. Analysis will dominate the scene only in the next section, where we concentrate on the role of biology. Knowledge of the limitations of biology is obviously important in the context of medicine simply because medical science is still mostly biology.

3. THE MENTAL AND THE PHYSICAL: FIVE PHILOSOPHICAL VIEWS

3.1. Psychophysical parallelism revived: R. Miller

Mind-body theorists, nowadays, will almost unanimously reject one form of dualism, which is called psychophysical parallelism. In its original form (as elaborated by Leibniz and Geulincx), it assumed that there is a close *correspondence* between mental events and physical events although there is no interaction between them. The synchrony of two clocks running "in parallel" was regarded as an appropriate metaphor. Thus parallelism is characterized by two major assumptions. It is committed to substantial differences between the mental and the physical, and it does not allow for causal relations between the two domains.

However, grand philosophical theories have a habit of surviving even long periods of dormancy. Miller (1981) has revitalized psychophysical parallelism by integrating recent philosophy, psychology and biology, his main emphasis being on neurobiology. He does not postulate a very close correspondence between the mental and the physical. But he accepts the basic tenets of parallelism.

Miller actually does not intend to give a full-fledged *philosophical* defense of parallelism. He is sceptical about the force of any philosophical defense. "The primary aim of this initial chapter is to present a brief discussion of these issues, and express some personal opinions about them. These opinions are hardly likely to be universally acceptable, and indeed it seems improbable that unambiguous resolution of the metaphysical issues will ever come about" (p. 7). The theory itself will have to bear the burden of proof. None the less, we will critically comment on Miller's introduction (pp. 10-19) of the two major assumptions of parallelism, dualism and absence of causal interaction, because he does present philosophical arguments.

The dualism premiss is introduced as follows.

We must reckon with two utterly different types of reality. This duality may be expressed in various ways. We may talk of the duality between external 'things', and the 'something' we are aware of internally (our 'mind' or 'consciousness'). We may talk of the duality of two processes – 'observation' and 'experience', which we should always try to separate. We may talk, alternatively, of the duality of information – 'objective' information (such as the sentences in a book, or the sequence of nucleotides in a gene) and 'subjective' information (inner experience having a distinct, though abstact form). In any of these cases, the dualist premise will influence the definitions we give to key words, such as 'thing', 'process', and 'information', all of which will have to be defined in a way which embodies dualism (p. 10).

Miller subsequently gives a few examples to show that introspection yields subjective descriptions (individual assessments) which one must take seriously. Objective descriptions function as collective assessments. Miller argues that one cannot get rid of subjective phenomena although they can be made objective to some extent. "...there must always be a crucial subjective judgement in deciding the significance of a result, simply beause part of the essence of science is that it is a mental activity" (p. 13).

We think that this approach cannot yield an adequate defense of dualism as a view "of the world", i.e. dualism as an *ontological* (Miller's term: metaphysical) theory. Miller only seems to have made some *epistemological* points (cf. duality of observation and experience, duality of information) which need not entail the ontological dualism which he ultimately wants to defend (cf. duality of things).

The concept of causality is introduced, rather bluntly, as follows. "... I venture to assume, at least, that a causal relationship is a relation between *observable* features of reality. According to this premise the relationship between the world of internal experience and the observable world can never be one of causality" (p. 14). This thesis, like the premiss of dualism, is presented without much defense. As it stands, it seems ambiguous. "Assume" suggests that Miller is addressing empirical issues, but the thesis may as well be interpreted as a covert *definition* of "causality". Thus Miller runs the risk of making parallelism true by convention.

Miller very briefly illustrates his view of causality with examples, one of which involves parapsychology.

... in discussions of parapsychology it is often unclear whether the causation is supposed to be a relationship between observables, or a relationship between an unobservable inner reality and an observable outer reality. Nevertheless, it is clear that no scientific investigation of these phenomena is possible if investigators adhere to the latter notion of causation. ... This is not to deny the subjective reality of paranormal experiences, merely to point out that any such experience in itself leads to a hypothesis about existence of parapsychological processes, rather than to a demonstration of its veracity (p. 15).

Examples of this kind will hardly help us to develop a theory of causality. Perhaps they are best regarded as a piece of methodological advice. If the notion of causality is given a wide meaning one will be left with untestable statements, so one had better restrict the meaning of "cause". This move may be sensible, but it does not yield much metaphysics.

Miller ends his introductory chapter by outlining how the two assumptions of parallelism bear on relations between mind and brain.

If we think we can speak of reality at all, we have a right to speak of both subjective reality (an individual view) and objective reality (a collective view). ... It is when we turn our gaze to the human brain that the two viewpoints are juxtaposed most starkly: for at once we must reckon not only with the objective facts of structure and causal relationship in the brain we are observing as experimenters, but also the subjective fact of inner experience in our own brain. It is at this point that our philosophy crystallizes. Somehow we must be able to reckon with the relationship between these two facets of our nature, so that it becomes valid to seek features in the anatomy or physiology of the brain which match with subjective descriptions. In this process, the logic we employ will not be like that used in explaining causal relationships. What we observe in the brain is so utterly different from what we experience with the brain that any suggestion of causal interaction between the two would destroy our concept of causality. However, we may use a more abstract form of logic, a *logical transformation* without causal implications (p. 17).

Unlike classics endorsing parallelism (Leibniz, Geulincx), Miller does not maintain that the transformation involves any strict analogy between the mental and the physical. "The term 'parallelism' is used here in a much more flexible sense than this, to imply the sort of mapping of a 'field' of information that might be possible in a multidimensional space" (p. 17).

The remarks on causality in the passages just quoted are specific versions of the general statements on causality which we have already commented on. "Mapping" is a new element, which is to play a central role in Miller's theory. He roughly proceeds as follows. First of all, he considers psychological issues revealing that one has to reckon with the unity of consciousness (admittedly not an all-pervasive kind of unity). He then argues that parallelism demands that the brain, likewise, acts in an integrative fashion. Otherwise one could not really understand consciousness. For example, neurons in the brain must be part of a single network, an *omniconnected structure* (although some localization of function is to be allowed). Similarly, specific elements of man's psychological make-up must somehow correspond with anatomical and physiological features.

The major part of Miller's book is devoted to a detailed study of specific correspondences. We will first comment on his general argument concerning the unity of consciousness, and then consider a specific example of correspondence.

However, the most important of the psychological premises, concerning the subjective unity of consciousness ... cannot be accounted for simply by a punctate relationship between structure and function in the brain. An hypothetical cortical surface, entirely composed of many loci, independent except in so far as they have overlapping sensory inputs or supply overlapping motor groups, cannot account for the existence of the central 'I' to which all portions of information are subjectively available. ... Unless some other principle of organization can be suggested, the above argument leads only to an infinite inward spiral, a veritable *reductio ad absurdum* (p. 44-45).

This passage is characteristic of Miller's style of reasoning. He assumes that biological data *somehow* explain features of consciousness. Of course the explanation cannot be causal, but Miller does not elaborate any methodology of non-causal explanation. He obviously assumes that "mapping" relations between the mental and the physical (mind and brain) must satisfy certain conditions if the physical is to have explanatory force vis-a-vis the mental. But the conditions involved are not explicitly stated. It is indeed unclear how parallelism could specify such conditions. Specific mapping relations may be discovered by empirical research. Parallelism, however, is compatible with *any* set of principles that satisfies minimal requirements, e.g., of consistency. The data from psychology and biology which Miller presents may suit the purpose of confirming general ideas about correlations between the mental and the physical. They fall far short of confirming his version of parallelism.

These comments apply as well to specific examples given by Miller, as the following characteristic passage reveals.

A point which immediately arises from this model is that the hippocampus, in so far as it samples neocortical activity, rather than receives detailed representations of its activity will signify 'match' or 'mismatch' (i.e. familiarity or novelty) without representing all the details contained within the context. We see in this model therefore, a parallel of certain subjective phenomena, in which feelings of familiarity and strangeness may be experienced in isolation, that is, dissociated from the detailed subjective impressions to which these feelings usually apply (e.g. the *déjà vu* phenomenon, or the feelings of strangeness and familiarity elicited in Penfield's temporal lobe stimulation studies ... (p. 152).

Miller, in short, seems to argue that it is clear that there is some kind of correspondence between the mental and the physical (the mind and the brain). At the same time he conveys the impression that we cannot really understand any observed correlations. He has apparently shown that the data of science (more specifically, neurobiology) are *compatible* with parallelism. Correlations, unfortunately, are compatible with almost any mind-body theory. One will need additional arguments if one wants to defend any particular theory, arguments which are likely to involve philosophy. Miller, however, has refrained from presenting a detailed philosophical defense of his theory, because he thinks that philosophical arguments can never be wholly convincing. Precisely this is why we remain unconvinced.

3.2. *The interface of mind and matter: Popper & Eccles*

Dualism comes in many varieties. A particularly strong variant is defended by Popper, the philosopher, and Eccles, the neurophysiologist, in a joint book (Popper and Eccles, 1981). Most dualists of our day will regard the term "mind" (a substantive!) as a metaphor pointing to a realm of mental *properties*. Popper and Eccles, however, apparently regard self-conscious minds as real entities which causally interact with brains. So their *interactionism* seems to be a kind of *substance dualism*. (This dualism is embedded in the pluralism of Popper's three worlds theory which we will not discuss.)

Popper's part of the book aims to disprove mind-body theories which are at variance with his own view and to mount a defense of dualism in general philosophical terms. Eccles gives an overview of neurophysiology (of the brain) as a basis for their mind-body theory.

We will concentrate on Eccles' contribution because the present section is specifically concerned with the role of biology. However, one should notice that the notion of *conjecture,* which plays an important role in passages we will analyse, is borrowed from Popper's philosophy. Popper has always argued that science must use the method of bold conjectures followed by critical tests (see especially Popper, 1963; some comments were given in chapter III, sections 2 and 3).

Popper and Eccles conjecture that the mind and the brain interact in a particular area of the left, dominant hemisphere of the brain, the "liaison brain". We will scrutinize, by a running commentary, Eccles' elaboration of this conjecture (quotations except the last one represent successive passages on pp. 362-363).

The self-conscious mind is actively engaged in reading out from the multitude of active centres at the highest level of brain activity, namely the liaison areas of the dominant cerebral hemisphere. The self-conscious mind selects from these centres according to attention, and from moment to moment integrates its selection to give unity even to the most transient experiences. Furthermore the self-conscious mind acts upon these neural centres modifying the dynamic spatiotemporal patterns of the neural events. Thus we propose that the self-conscious mind exercises a superior interpretative and controlling role upon the neural events.

"The self-conscious mind ... give[s] unity ... to ... experiences." The last part of this statement can be interpreted in two very different ways. One, there is experienced unity. Two, there is something which has unity and which is experienced. The first interpretation makes the (second part of the) statement relatively innocuous. But the second interpretation would make it problematic. In the last few decades, psychologists and philosophers of psychology have shown that the term "consciousness" has many different meanings. Furthermore, they have argued that results of psychological research suggest that there is no "unity" of consciousness (on any inter-pretation of consciousness) *although* there is *experienced* unity (for sources, and some comments, see section 4). Perhaps they are wrong, the issue is contentious. But it should be clear that one cannot simply disregard (philosophy of) psychology if one wants to elaborate hypotheses about consciousness. Eccles is apparently unaware of publications dealing with the foundations of psychology.

The ambigous formulation used by Eccles facilitates the amalga-mation of two different statements (cf. the interpretations mentioned). This easily results in fallacious reasoning, from *experienced* unity to the exist-ence of a "self-conscious mind" (unity which is experienced). True, Eccles does not *explicitly* reason in this way. One is forced to guess at his ar-guments precisely because his wording is ambiguous.

A key component of the hypothesis is that the unity of conscious experience is provided by the self-conscious mind and not by the neural machinery of the liaison areas of the cerebral hemisphere. ... Our present hypothesis regards the neuronal machinery as a multiplex of radiating and receiving structures: *the experienced unity comes, not from a neurophysiological synthesis, but from the proposed integrating character of the self-conscious mind.* ...

Eccles does seem to have slipped here to the second interpretation. The "unity of conscious experience" (something which has unity and which is experienced?) must be explained, and Eccles has a hypothesis (which he shares with Popper) that may provide an explanation. However, the phrasing in the hypothesis is different. Here we meet "experienced unity" again (two interpretations possible; see above). Many psychologists and philosophers of psychology would argue that man is *deceived* by ordinary "experience": there need not be any real unity.

In refining the conjecture we have to imagine that in the liaison areas of the cerebral hemisphere some sensory input causes here and there an immense ongoing dynamic pattern of neural activity. ... the primary sensory areas project to secondary and these to tertiary and so on ... . In these further stages the different sensory modalities project to common areas, the polymodal areas. In these areas most varied and wide-ranging information is being processed in the unitary components, the modules of the cerebral cortex ... .

"*In* refining the conjecture" indeed. One has to read carefully here. Eccles is *not* presenting a refinement of the conjecture. The neurobiological *part* of the conjecture is refined, the mental remains elusive. Now Eccles is a brilliant neurobiologist. But an elaboration of neurobiology only will not clarify relations between biological processes *and the mental.*

We may ask how is this to be selected from and put together to give the unity and the relative simplicity of our conscious experience from moment to moment? As an answer to this question it is proposed that the self-conscious mind plays through the whole liaison brain in a selective and unifying manner.

The question is not really answered (cf. "how"!). Eccles states *that* the mind plays a role, not *how* it plays its role.

An analogy is proved by a searchlight ... . Perhaps a better analogy would be some multiple scanning and probing device that reads out from and selects from the immense and diverse patterns of activity in the cerebral cortex and integrates these selected components, so organizing them into the unity of conscious experience.

Analogies may indeed be useful.

Thus we conjecture that the self-conscious mind is scanning the modular activities in the liaison areas of the cerebral cortex ... . From moment to moment it is selecting modules according to its interest, the phenomenon of attention, and is itself integrating from all this diversity to give the unified conscious experience. Available for this read-out, if we may call it so, is the whole range of performance of those areas of the dominant hemisphere which have linguistic and ideational performance or which have polymodal inputs. Collectively we will call them *liaison areas* .

"*Thus* we conjecture that the mind .. *is* scanning." The wording could suggest that Eccles is presenting an argument. But there is no argument. The analogy is still an analogy only.

We do not think that this kind of reasoning will make one understand *how* the mind and the brain interact. Eccles gives many examples of specific associations between the mental (however conceived) and the physical (anatomy and physiology of the brain). *If* one assumes that the mental and the physical are in essentially different realms, the evidence does indicate that the realms interact. But it hardly shows *how* they interact.

In the main text of the book, which precedes a discussion between Eccles and Popper, Eccles only refers to the mechanism of interaction in rather vague terms. His discussion of Kornhuber's experiments on action potentials in the cerebral cortex associated with willed action is an example. The action potentials are preceded by a small so-called readiness potential.

It can be presumed that during the readiness potential there is a developing specificity of the patterned impulse discharges in neurones so that eventually there are activated the pyramidal cells in the correct motor cortical area ... for bringing about the required movement. The readiness potential can be regarded as the neuronal consequence of the voluntary command (pp. 284-285).

How does the non-material mind recruit the energy required for generating a readiness potential and other neural phenomena? The question is not answered in the main text of the book. In the appended discussions, however, the authors (especially Popper) show their hand with astonishing naivety.

Popper had already suggested (p. 180) that Descartes' view, though crude, was on the right track. Descartes postulated a mechanical interplay between the mind, located in the epiphysis, and corpuscles (spiritus animalia) in the brain cavities. This conception proved to be unacceptable because it is hardly compatible with conservation laws. However, if one replaces Descartes' mechanical phenomena by electrical ones, one gets an attractive view. The energy which the mind requires to deflect electrical currents "is almost equal to zero". So one need not bother about conservation. On pp. 539-547 Popper elaborates this view. He offers the following arguments.

1. Conceptions of the material world as a closed system have repeatedly failed. Newton's mechanical world was invaded by electrical forces and the ensuing new system lost its closure when nuclear forces were discovered. Such changes always required revisions of the conservation laws which characterize closed systems. Therefore, one should not worry too much about the objection that dualism violates the law of energy conservation. Further developments in physics will smooth out the discrepancy.

2. Moreover, new physical laws are not even strictly necessary. The evidence plainly shows that the mental and the physical interact. How then could one fail to assume that the material world is open to the world of the mind and *vice versa*?

3. The second law of thermodynamics is not really an obstacle. The law is compatible with interactions between the mind and the brain because it is statistical. Local temporary decreases of entropy induced by the mind will be compensated by the brain becoming tired or hot, i.e. by a degradation of the energy which restores the entropy balance.

4. Finally, why bother about such explanations at all? The energy required by the mind to set the brain to work may· be unmeasurable. Moreover, the brain has ample possibilities to regulate blood supply so that areas mobilized by the mind get additional oxygen (Eccles gives "evidence" on pp. 545-546).

We do not want to give detailed comments on these arguments. They are speculative, but speculations are sometimes healthy. They also represent a blatant *petitio principii*. Nowhere in their voluminous book have Popper and Eccles made interactions between the mind and the brain intelligible.

3.3. The eminence of matter: Bunge

Dualism, as the previous examples showed, has a difficult job to perform. It has to keep the mental and the physical at opposite sides of the fence, but it must also provide sensible connections. Will monism fare any better? Nowadays, monism mostly takes the form of materialism. Thus it somehow has to get rid of the mental. Mental events, so it is argued, are actually special physical (neurophysiological) events. Common sense and folk psychology (and much academic psychology) seem to indicate otherwise, but they are mistaken.

Bunge (1980, 1981) is among the philosophers who take this line. He defends a position which he calls *psychoneural identity theory*, or *emergentist materialism* (1981, p. 81). We will first consider one vital element in the theory as expounded in his 1980 book.

After a general outline of his approach, Bunge first introduces key notions and principles concerning brain functions. He then tackles mental states and processes. Right from the start, mental processes *as a general category* are explicitly *defined* in terms of specific physiological processes in the brain (1980, p. 74). We omit details of the definition because Bunge employs a mathematical formalism which one can only understand after going through a lot of technicalities earlier introduced by him. What matters here is *that* the mental is given a definition in physiological terms. Bunge subsequently shows that the definition of the mental in these terms has far-reaching implications ("corollaries"), e.g.: "All mental disorders are neural disorders" (p. 75).

How could such an approach be justified? Consider the following remarks which illustrate Bunge's stratagem.

We have not characterized mental states and events independently of brain states and events, as is usually done. We have not proceeded this way for a number of reasons. Firstly, because mentalistic predicates, such as "sees red" and "thinks hard" are, though indispensable, coarse and vulgar rather than exact and scientific. Secondly, because the whole point of the neurophysiological approach to mind should be to get away from the ordinary knowledge approach and render mind accessible to science. Thirdly, because, if mental events are characterized independently of brain events, then the identity theory turns out to be either idle or false ... . Fourthly, because some physical events too might be given (simple-minded) descriptions in everyday mentalistic language, and so would be "proved" to be mental (p. 80).

Bunge is right, of course, in assuming that common sense and folk psychology have their limitations. But would it be helpful to *define* "seeing red" and "thinking hard" in physiological terms? We think that this is an awkward move. If materialism is made true by definition, there cannot be factual evidence for or against it!

However, Bunge would definitely want to *test* his theory. Consider how he proceeds in commenting on "consciousness" (p. 175). He first gives the following definition. An animal b is conscious of a brain process x in b iff b thinks of x. Then he postulates "that a conscious event is a brain activity consisting in monitoring ... some other brain activity". That is, conscious events are activities of neural systems. As yet the nature of these systems is unknown.

We doubt whether the definition is an explication of any common usage of the term "consciousness". Mentalists would argue that the *contents* of consciousness do not represent brain processes at all. Bunge's concept of consciousness, however, seems to imply that they do represent such processes (cf. "a brain process x" in the definition). Thus his concept apparently presupposes "psychoneural identity", the very hypothesis he should test. *Being* conscious ("of brain process x"), likewise, is almost *defined* in neurobiological terms, because "thinking" is apparently portrayed as a neuriobiological phenomenon (chapter 7). Bunge seems to be running around in circles.

At first sight, in short, Bunge's strategy is contradictory. He seems to define the mental in terms of the physical *and* to investigate factual relations between the mental and the physical. One cannot do such things at the same time! However, the situation is not that simple. Bunge would reply that mental phenomena are physical phenomena of a special kind, and that he is investigating factual relations between two categories of physical systems. Would that reply be adequate? That depends. In view of Bunge's strategy, which involves an important place for science, one would want to confront his theses with real science. However, Bunge is not offering much in the way of real science. He often *refers* to results of science which tend to confirm his views in very general terms. But right from the start, he puts them in an abstract, idiosyncratic framework of his own making. Moreover, he often argues that *future* science will deal with apparently plausible theses which contradict his views. This makes it almost impossible to refute his materialism. We can but conclude that it is a form of sheer dogmatism.

Consider the following example which reveals how Bunge immunizes his views. He has defined the mental *as a general category* in physiological terms, but he cannot do that for all *specific* mentalistic concepts which are commonly used. No matter. He simply admits that *current* neuroscience is not a sufficient basis for understanding the mental. Why should that be so?

One reason is that psychology contains certain concepts and statements that are not to be found in today's neuroscience. Consequently neuroscience must be enriched with some such constructs if it is to yield the known psychological regularities and, *a fortiori*, the new ones we would like to know (p. 214).

Current explanations of mental phenomena in Bunge's style will obviously have to be very sketchy. And so they are. Consider. On pp. 212-213 Bunge gives a succinct survey of "mental properties, events, and processes in man", and of rival accounts to explain them (two alternatives, a mentalist account and a psychobiological account which he himself favours). We will mention two items without further comments. "Why is *z* in a happy mood?" Mentalist account: "because her mind is in a state of happiness". Psychobiological account: "because her brain is brimming with biogenic amines". "How did *z* recover from his depression?" Mentalist account: none. Psychobiological account: "*z* was given an antidepressant drug which prevents the breakdown of norepinephrine".

In a later publication (1981), Bunge articulates his scientific materialism by arguing that it need not take the form of reductionistic descriptions of life in terms of physics. Science beyond physics has continually incorporated new concepts which describe qualitative transformations of matter during evolution. As molecules form cells and cells form organisms, matter develops new structures and properties. "A materialistic conception of life has got to acknowledge emergence, i.e. the fact that systems possess properties absent from their components" (1981, p. 8).

Consciousness is an emergent property resulting from a qualitative transformation of (part of) the brain during human evolution. No immaterial principle is needed to explain this evolution once one rejects the antiquated static conception of matter. Modern emergentist materialism, unlike reductive materialism, regards matter as qualitatively changeable. It allows for a hierarchy of systems (cells, organisms, social systems, cultural systems). Systems at each level of the hierarchy have properties which their components lack (1981, pp. 26-31).

The development of conscious mental states in higher animals exemplifies the emergence of new properties. Early nervous systems are committed to the regulation of the internal environment, to reflex activity and other preprogrammed functions. Higher animals, however, have additional neural systems developped later which are committed to mental functions with more plasticity such as learning. These systems, "psychons", jointly constitute the "psychosystem" (the mind). Psychon-functions, being neural processes, can of course influence other somatic functions and *vice versa*. So mind-body interactions are not problematic anymore (1981, p. 26).

Could one *explain* emergent properties like conscious mental states in neurobiological terms? Bunge seems to be saying that it can be done *in the future*, in terms of a neuroscience which still has to be enriched with new (neurobiological) concepts for the mental. Meanwhile one will obviously have to be content with disjunct theories, of neurobiology and psychology.

3.4. Mentalism in new garbs: Sperry

Sperry is doubtless among the best researchers of the brain. His research on split-brain patients has had a marked influence on various fields of science. We will not discuss this work, but concentrate on his philosophical views on the mental and the physical as explained in Sperry (1969, 1976, 1983).

Sperry emphatically rejects dualism. Statements to this effect, however, must be qualified since he reserves the term dualism for substance dualism as defended by Popper and Eccles (1981). Other mind-body theorists would (rightly) call him a property dualist. Sperry at any rate wants to give the mental a status of its own and he ascribes causal efficacy to mental events. Let us see how he proceeds.

Sperry, like Bunge, gives the concepts of emergence and system important roles. Phenomena of inner mental experience are phenomena in their own right. "Rather than being identical to the neural events, as is generally understood, they are emergents of these events" (1976, pp. 166-167). Their properties are *pattern properties* with a reality (and a causal role) of their own, beyond the summed properties of constituents.

The reductionist approach that would always explain the whole in terms of the parts leads to an infinite regress in which eventually everything is held to be explainable in terms of essentially nothing. ... Let me repeat that the thing to remember in this connection is that, in causal interplay between systems and their surroundings, the spatial and temporal relationships of the constituent parts of a system have in themselves important causal efficacy over and above the properties of the parts *per se* (1976, p. 167).

This passage suggests that Sperry would endorse emergentist materialism. However, he emphasizes that we cannot now explain processes at the level of consciousness. Adequate explanation will require "a new technology that will enable us to record the pattern aspects of cerebral function which at present can only be extrapolated from indirect or highly particulate sampling procedures" (1969, p. 535).

Sperry's 1983 book shows that he is not really a materialist. He seems to realize that inner experience and physiological data can hardly be accomodated in a single theoretical framework. As a result, his views tend to become inconsistent.

To make a long story short, if one keeps climbing upward in the chain of command within the brain, one finds at the very top those overall organizational forces and dynamic properties of the large patterns of cerebral excitation that are correlated with mental states or psychic activity. [Sperry continues with an example concerning pain "in" a phantom limb, i.e. pain which seems to "reside" in an amputated limb.]

In regard to the pain in a phantom limb, my contention is that any groans it may elicit from our patient and any other response measures or behavioral outputs that may be taken to be the result of the pain sensation are indeed caused not by the biophysics, chemistry or physiology of the cerebral nerve impulses as such, but by the pain quality, the pain property, per se. This brings us, then, to the real crux of the argument. Nerve excitations are just as common to pleasure, of course, as to pain, and the same is true of any other sensation. What is critical is that unique patterning of cerebral excitation that produces pain instead of something else. It is the overall functional property of this pain pattern as a pattern that is critical in the causal sequence of brain affairs. This pattern has a dynamic entity, the qualitative effect of which must be conceived functionally and operationally and in terms of its impact on a living, unanesthetized cerebral system. It is this overall pattern effect in brain dynamics that is the pain quality of inner experience. To try to explain the pain pattern or any other mental qualities only in terms of the spatiotemporal arrangement of nerve impulses, without reference to the mental properties and the mental qualities themselves, would be as formidable as trying to describe any of the endless variety of complex molecular reactions known to biochemistry wholly in terms of the properties of electrons, protons and neutrons and their subnuclear particles, plus (and this, of course, is critical) their spatiotemporal relationships. By including the spatiotemporal relations, such a description becomes feasible in theory, probably, but fantastically impractical. Moreover, by the time science arrives at a point where it can describe the critical details of the impulse pattern of a mental experience in the functional terms and setting required, it will be describing, in effect, the conscious force or property itself. When we reach such a point, the conscious force will be recognized as such, and we will be calling it just that – or at least that is the hypothesis I am putting forward (1983, pp. 33-35).

Let us scrutinize this passage to uncover the relations between the mental and the physical that Sperry postulates. At the outset, he seems to suggest that mental events (exemplified by pain sensations) and physical events are fundamentally different: patient responses are not caused by biophysics etc., *but* by the pain quality as such. Subsequent comments on "patterning" carry the same suggestion. The last part of the passage, however, strikes a different note. In theory (though not in practice) one could describe inner experience in physical (biological) terms! If one were to succeed in doing this, however, one would be describing the "conscious force or property itself". Sperry here seems to postulate that mental events are identical with physical events (of a sophisticated sort) in the brain.

Sperry, in short, solves mind-body puzzles by having his cake and eating it. Mental events and physical events are quite different. Sperry honours this item by introducing a form of property dualism. It is difficult to understand interactions between the two kinds of events if they are kept in different realms. Sperry grants this as well, and then seems to solve the problem by surreptitiously introducing property monism. If this is his solution of the mind-body problem, it is no more than a disappearance trick.

3.5. The reality of systems: Von Bertalanffy

The views developed by Bunge and Sperry resemble those of the Austrian biologist Von Bertalanffy in various ways. Therefore his views need to be mentioned here although he does not belong to the Anglo-Saxon tradition. Von Bertalanffy is famous for his growth equations and other contributions to theoretical biology. Here we will consider his general "philosophy".

Von Bertalanffy maintains that the mind-body problem should be approached scientifically rather than philosophically. "What were once considered philosophical problems of epistemology and metaphysics have become empirical questions to be investigated by scientific methods" (1964, pp. 30-31). The dichotomy of the mental and the physical is not a very fundamental feature of reality. We have *chosen* in our culture for a mind-body antithesis which engendered physics and psychology as conceptual constructs representing certain aspects of reality. (This thesis resembles views defended by phenomenologists; see section 5; in other respects Von Bertalanffy is definitely not a phenomenologist.) What consequences does this have?

The first consequence is that we have to relinquish so-called reductionism. The concepts of psychology cannot be reduced to those of neurophysiology. Neither is the mental world an epiphenomenon to the physical world. ... Both the worlds of physics and of psychology are constructs to bring certain aspects of the experienced universe under the rule of law" (1964, pp. 40-41).

Von Bertalanffy subsequently claims: "Nevertheless, excluding reduction of psychology to neurophysiology, we can indicate what their relation is and how unification of both fields may be sought". The first step towards this unification is the postulation of an isomorphism between the constructs of psychology and neurophysiology. This isomorphism does not imply simple similarities but rather such relations as between a code and the encoded; it is but a minimum requirement which must be met "because neurophysiology would make no sense without correspondence to mental processes ..." (p. 41).

Unification of physiology and psychology is to be realized by "constructs which are generalized with respect to physics and psychology" (p. 41). What would a generalized theory look like? "Recent theory construction in cybernetics, information theory, general systems theory, game and decision theory, etc., elaborates constructs precisely of this kind - that is, constructs that are neither physical nor physiological, but are applicable to both fields" (p. 41).

General systems theory, which was developed by Von Bertalanffy himself, plays an important role (for a survey see Von Bertalanffy, 1968; the theory was later expanded by Miller, 1978). In an earlier book (1949), he described it as the mathematical formalization of relations and principles which apply to all systems, irrespective of the nature of their components or relations between components. There are systems at various levels of organization (biological, psychological, social), which combine to form hierarchies (1949, chapter II; 1964, p. 33).

Von Bertalanffy's general theory is in some respects ambiguous. The emphasis on system hierarchies can be interpreted as an attempt to devise methods for a *substantive* integration of disciplines. New formal approaches could help us to achieve this goal. On this interpretation, Von Bertalanffy's strategy is a sound one. In retrospect, however, one can only conclude that no grand integration has been forthcoming in spite of later attempts to develop the program in greater detail.

The theory can also be regarded as an attempt to establish *methodological* unity among the sciences. If that is its purpose, one must realize that the sciences share *many* useful methods. Calculus, elementary logic, indeed almost any formal method may be useful.The unqualified suggestion that application of any particular method will result in a unified science is misleading.

There is nowadays a useful mathematical systems theory, which is not associated with any unity of science ideal. General systems theory is different. It is often presented as a comprehensive philosophy which reveals the true nature of reality. We doubt whether Von Bertalanffy would have liked the turn his theory has taken!

In the philosophy of medicine, "systems theory" is mostly associated with "holism". The holism movement will be discussed in chapter VI.

3.6. Afterthoughts

How much has biology helped us? *Miller* showed that neurobiology is *compatible* with psychophysical parallelism. But it is compatible with other mind-body theories as well. To make a choice, one needs good *philosophical* arguments which Miller does not want to elaborate as he distrusts the force of philosophy. Eccles, with Popper, postulated interaction between the mental and the physical. He only elaborated neurobiological *aspects* of the interaction. The nature of the interaction, therefore, remained elusive. *Bunge* almost made his view true by definition. To confirm it we would need future science. *Sperry* also seemed to rely on future science. Moreover, his philosophical position remains unclear. *Von Bertalanffy* likewise thinks in terms of future developments in science.

The philosophy we uncovered in the writings of these authors is mostly kept at the margin. It apparently represents classical positions which have been defended throughout the ages. Biology did not seem to offer any new arguments. But perhaps one had better turn to psychology, the science of the mental, for new inspiration.

4. A SCIENCE OF THE MENTAL?

4.1. Idiosyncracies of the mental

Philosophy inspired by mere biology will not be very helpful in attempts to solve the mind-body problem. Would psychology, the science of the mental, be a better source of inspiration? Scientific psychology should indeed be helpful, but it faces problems of its own. It has long kept consciousness (defined in terms of inner experience) outside its subject matter. Methodology was thought to forbid reference to introspection, so behaviour got a central place. The organism (animal or man) was regarded as a black box, which one could understand by the study of stimuli and responses. Classical behaviourism, with its materialistic outlook, was eventually challenged by the "cognitive revolution" which made the study of internal states respectable again. Even modern *cognitive* psychology, however, is apparently compatible with materialism. For example, its dominant variant, functionalism, often interprets internal states in terms of physical processes (cf. the computer metaphor). So consciousness still has to fight to get a place in the subject matter of psychology (see e.g. the volumes edited by Underwood and Stevens, 1979, 1981, and Underwood, 1982). Behaviourism, for that matter, is not quite dead. Rachlin (1985), e.g., has recently defended a behaviouristic account of pain (see also Zuriff, 1985, and Natsoulas, 1986, for surveys of modern behaviourism). Rachlin's account of various theories of pain illustrates how divided opinions over the mental are in psychology.

Various thorny philosophical problems plague the foundations of theoretical psychology. One of them concerns the explanation of action.

Let us give the mental a status of its own, and try to mount explanations in mentalistic terms. Goals, intentions, motives, reasons, and the like are obvious items to consider. Could one give a causal explanation of any particular action in such terms? Our writing this is *motivated* by the *intention* of convincing you. We are writing since we *believe* that you may become convinced. Is our intention a *cause* of the writing? There are various problems here. It might be argued that writing (in the absence of countervailing factors) is simply a criterion for the presence of a writing-intention. That is, one may regard the general statement "Whenever person P (in the absence of countervailing factors) intends to do X, he will do X" as a tautology. Malcolm (1968), in a famous article, has given a detailed analysis along these lines. If he would be right, we would not be left with any substantial explanation. Tautologies cannot represent causal relationships.

However, if Malcolm is wrong, our task will not be easy either. Perhaps intentions, reasons, etc. are causes. If that is true, we will want to know *how* they result in any action, and that is a baffling matter. Should it be possible at all for psychology to explain actions? The question has been answered in many different ways (see e.g. Cummins, 1983; Flanagan, 1984; Moore, 1984; Russell, 1984; Smith and Jones, 1986).

Russell (1984), in a lucid survey, analyses the situation as follows. If the connections between intentions and actions are so tight that ordinary explanations are ruled out, psychology will tend to become a *hermeneutical* science (no causal explanation, explanation as description). If the ties are loosened, so that intentions may figure as causes, one will still not have any laws for explaining actions in terms of intentions. Psychology will then be an *ideographic* science (a science without laws). The dilemma, however, arises only if we attempt to explain *individual* actions, behavioural episodes, or mental states. Such attempts belong to folk psychology. Academic psychology, according to Russell, is concerned with a more general issue, viz. the explanation of mental competence in the human species (for further reference, see also Wilkes, 1980).

Other methodological problems are generated by the use of mentalistic *language*. Let us concentrate on an example. Sentences like "*X believes* that Maggy is wrong" turn out to be rather peculiar when we try to apply substitution. Suppose Maggy is in fact the queen of the U.K., although X doesn't know that. By ordinary substitution, this would give "*X believes* that the queen of the U.K. is wrong". However, this will not do. If X believes that Maggy is wrong he need not at all believe that the queen is wrong. Substitution is not allowed because the context is *intensional*. "Believe", "know", "desire", in short, predicates denoting propositional attitudes will always generate this problem. Now, the language of natural science is extensional, or at least it is supposed to have this feature. This alone suggests that it may not be easy to integrate (mentalistic) psychology and biology in order to understand relations between the mental and the physical. Various philosophers (e.g. Heil, 1986) have indeed suggested that psychology cannot be continuous with biology.

The characteristics of mentalistic predicates have long been thought to support dualism. Thus one could argue that expressions like "X believes p" and "X has brain state B" are so different that they can hardly refer to the same kind of event. However, *this* argument is not accepted anymore. Linguistic peculiarities are a poor basis for inferring what the world is like. Many functionalists (e.g. Dennett, 1985) happily combine mentalistic language with a materialistic view.

Functionalism regards mental states as identical with functional states of the brain, but it does not try to integrate biological and psychological theory. It is not easy to study cognitive processes "as they occur in the brain". What matters for functionalists is how such processes are organized. This issue can be approached quite well with computer models. For example, thinking is regarded as a computational process.

We are rather skeptical of functionalism. It can hardly deal with essential characteristics of the mental such as qualitative content of consciousness, the subject-object distinction presupposed by knowledge of the world, etc. (see Russell, 1984). Psychology in the hands of functionalism again appears to loose what should be a primary datum of its subject matter, consciousness.

Problems like those mentioned above keep cropping up in discussions on the *foundations* of psychology. Perhaps they should primarily concern the philosopher, not the psychologist. Let us try to forget about them for a moment, and consider what happens in mundane psychology.

The phenomena studied in psychological research belong to categories like learning, memory, emotion. It is obvious that research will give one good theories only if one manages to elaborate a neat system of categories. Unfortunately, it is equally obvious that the systems now available are anything but neat. For example, "learning" represents an array of quite different processes. Psychologists, whatever their persuasion, agree about this. Could one subdivide the category of "learning, such that one gets concepts for (more specific) "unitary" phenomena? Perhaps so. At the moment, however, there is no agreement on the value of any particular subdivision. The same goes for almost all the major categories in the domain of psychology, first and foremost *consciousness* (see e.g. Wilkes, 1984).

For example, consider visual perception, or *seeing*. One would think that statements with the form "X sees object Y" are fairly unproblematic, and that "seeing" is a neat category. Likewise, "ability to see" *versus* "blindness" looks like a sensible dichotomy. But the phenomenon called *blindsight* disarrays common distinctions. Weiskrantz et al. (1974) discovered it in experiments with a patient (referred to as D.B.) who underwent brain surgery. D.B. reported that part of his visual field had disappeared. When an object was placed outside his remaining field of vision, he was unable to recognize it. However, the experiments showed that he was able to identify the object by *guessing*. Was D.B. *seeing* the object? In one sense of the term, he was. In another sense, he definitely did not see it.

How should one interpret the phenomenon? There are various possibilities. Seeing is normally accompanied with consciousness. Perhaps D.B.'s "seeing" is an example of seeing without consciousness. Alternatively, it could be argued that *conscious* seeing normally goes with self-consciousness. When one sees an object one also knows *that* one sees the object. On this view, D.B. was seeing objects without knowing it. Whatever view one takes, visual perception cannot be regarded anymore as a unitary phenomenon (for futher comments see Weiskrantz, 1987).

The point of the example is not merely that theories of common sense or *folk psychology* are problematic. There are problems for scientific psychology as well because one needs folk psychology as a data base.

As matters now stand, philosophy of psychology and psychology are not sufficiently well-equipped to solve any mind-body problem. But perhaps solutions will be forthcoming when they are combined with other disciplines. The next section will deal with some recent attempts to develop more integrative views.

4.2. The failure of integrationism

Will science help to solve the mind-body problem? Biology alone is not enough. The authors discussed in section 3 gave biology an important role. They almost *disregarded* problems which psychology has uncovered. Perhaps that is what made them fail. Psychology is not enough either.

Would an approach which integrates various fields of science put the mind-body problem in a better perspective?

Biology (more specifically, neurobiology) and psychology (more generally, cognitive science, a combination of cognitive psychology and various other fields) are now separeted by a wide gap. But there are attempts to integrate them (see e.g. LeDoux and Hirst, 1986). Will integration be helpful? We are sceptical. The ideal of a substantively unified science, popular in the heydays of logical positivism, is not realistic. Theories will not remain coherent if they are taxed beyond certain limits. For certain purposes, it is wise to put physiological theories of perception in an ecological perspective. In a different context, anatomy will be relevant as well. But attempts to combine all the areas of, say, biology and psychology at the same time seems to be misguided.

Meanwhile, there is an ever-increasing tendency to integrate disciplines which seems to reflect a revival of the unity of science ideal. The result is staggering. There is psychoneuroendocrinology, psychoneuroimmunology and what not. The result is more disciplines, not less.

We refuse to believe that all this will lead to a solution of any mind-body problem. What is loosely called mind is covered by one cluster of disciplines with many connections between them. The body is covered by a different cluster. The two clusters don't mix. So there is a gap between the mental and the physical. Of course the clusters may come to be rearranged so that one gets other major gaps. Then the mind-body problem will just take a different form.

We will comment on two attempts to integrate various fields of science *and* philosophy which bear on the mind-body problem (Wilkes, 1980; Patricia Churchland, 1986).

Wilkes (1980), in a provocative article, first of all notes that philosophers have been very one-sided in their approach of the mind-body problem. They "have always interested themselves primarily in the relations that hold, or fail to hold, between the mental and the physical. ... Strangely enough a prior question is usually ignored, namely just what are these mental and physical relata, the items that provide the two terms of the relation?" (p. 111). Mental relata have received *some* attention, but physical relata are hardly discussed at all. Philosophers apparently do not want to trespass in the garden of science. Wrongly so.

We can but agree. For example, if one skims Anglo-Saxon anthologies dealing with the mind-body problem which appeared in the sixties and seventies (Hook, 1960; O'Connor, 1969; Borst, 1970; Rosenthal, 1971), one is apt to feel lost in a world without contact with either science or common sense. But the tides are turning (cf. examples in section 3).

Wilkes does not aim at defending a particular mind-body theory. She is concerned with conditions which such theories must generally satisfy. Any integrative view of the mental and the physical will have to presuppose lawlike correlations between mental states and physical states. This implies that there are constraints upon the relata. First of all, one needs sensible classifications for both kinds of states.

One avenue had better be blocked right from the start. Wilkes argues that there cannot be lawlike correlations between brain states specified in terms of the behaviour of individual neurons taken together, and mental states such as remembering one's need for toothpaste. She convincingly shows that "there can be no worthwhile correlations between specific thoughts and brain states" (p. 116). One clearly needs items at much higher levels of generality (see also section 4.1). It will not be easy to develop good classifications for such levels. Firstly, localization theories which associate mental functions with particular areas in the brain have been disproved. So physical relata will have to be complex. Secondly, the prevailing categories of psychology for mental relata are very unsatisfactory (see also section 4.1).

At this point, Wilkes' arguments take an interesting turn. How could psychology arrive at a more satisfactory classification? Time and again, classifications have been overturned by *physiological* investigations. Psychology needs biology to structure its own domain. Thus the need of correlations between the mental and the physical gets the function of a constraint upon theory construction. "...so the two sciences [psychology and physiology] beg the question of the mind-body problem in advance" (p. 127).

We have one comment. If the integration envisaged by Wilkes would materialize, one would have some new science, but the question begged would *ipse facto* not be answered.

Even if lawlike correlations between the mental and the physical would be established without question begging, the mind-body problem would not yet be solved. Correlations are compatible with many mind-body theories.

Patricia Churchland's book "Neurophilosophy, toward a unified science of the mind/brain" (1986) seems to represent the best integrative approach of the mind-body problem now available. It gives an excellent survey of modern neurobiology, philosophy of science, and psychology and cognitive science. Functionalism and (more generally) cognitive science, on her view, are a poor basis for a unified science of the mind/brain, because they disregard neurobiology. She outlines a new, materialistic approach, developed together with Paul Churchland (1985), which accomodates much biology.

Churchland emphasizes the import of complex mapping relations which exist between various areas of the brain. Her survey amply illustrates the shortcomings of localization theories. Mapping relations should play an important role in integrative theories of the mind/brain (Churchland presents an outline for such a theory which we will not discuss).

The philosophical position taken by Churchland, *reductive* material-ism, calls for some explanation. Reduction in a classical sense of the term is a relatively straightforward logical relation between scientific theories. Theory *A* is reduced to theory *B* iff *A* can be deduced from *B* (see section 2). This straightforward species of reduction is hardly instantiated in science. Churchland envisages a more sophisticated kind of "reduction" which allows for complex relations between theories. In succinct terms, her version of reductive materialism amounts to the thesis that theories of psychology and neurobiology will co-evolve in such a way that the mental will ultimately be explained in biological terms.

Reduction is a relation between *theories*. That is an important specification. Statements to the effect that mental states or events can be reduced to physical states or events are best regarded as elliptical formulations. If one is discussing reductive relationships among states or events, one is doubtless putting them in a theoretical context (of ordinary science or folk science). So the point is whether *theories* about the mental are reducible to other theories.

At present there are no adequate psychological theories of the mental. Churchland, like Wilkes, thinks that neat classifications of mental events can only be arrived at through a biological approach. In this connection she argues that the unity of science ideal should have the force of a heuristic maxim. We have already argued that we would not accept *this* maxim.

Churchland admits that all the classic options in mind-body matters are still open *in principle*. However, are alternatives to materialism plausible? Strong dualism (substance dualism) does not seem plausible. Its adherents have consistently failed to explain *how* interactions across the ontological gulf it postulates are possible. Weak dualism (property dualism) is not in a very comfortable position either. It often appeals to *emergence*

(cf. various views in section 3). For example, it has been argued that consciousness cannot be explained in biological terms because it is an emergent
('irreducible") property.

Here Churchland argues that the dualists have failed to make their
case because they have disregarded the import of theories in discussing
emergence. States, processes, things and events cannot be emergent *as such*.
In more accurate terms, "emergence" applied to such items without qualification, is not a very clear notion. One needs *theories* about consciousness
in order to show that it is an emergent. And there are no adequate theories.

We tend to agree. The authors we discussed in section 3, monists and
dualists alike, exemplify the naivety which Churchland criticizes. They
simply disregarded psychological theory. In doing that, they implicitly
must have used theories of folk psychology which should not be taken for
granted.

However, Churchland's thesis that the mental *can* be accomodated by
materialism needs more defense as well. Her acceptance of materialism is
ultimately based on a *decision* to opt for a particular version of the unity of
science ideal *and* the *conviction* (or perhaps hope) that academic psychology and folk psychology will change in such a way that they will be
encompassed by biology. We ourselves do not want to make this decision.
And we are not convinced that psychology will follow the course set out by
Churchland.

Our impression remains that "integrations" in current science are
numerous indeed. So numerous, that desintegration may well become an
important feature of science *as a whole*.

The issues are clearly undecided. The science of the future cannot
help us now. But the science of our time does not have a solution either.
Would philosophy without science be a better bet? The next section will
address this question.

5. THE PRIMACY OF HUMAN EXISTENCE: PHENOMENOLOGY

5.1. Overview

So far, we have only considered recent Anglo-Saxon approaches to the
mind-body problem which put much emphasis on science. None of them

proved to be satisfactory. Would philosophical traditions which developed in continental Europe fare any better? We have chosen to concentrate on one school, phenomenology, which is now apparently past its prime (as a *philosophical* school, that is). Our choice was motivated by the following considerarations.

Phenomenology is a school which emphatically wants to deal with the mind-body problem. Other schools, e.g. structuralism, critical philosophy and hermeneutics have a different emphasis. Moreover, phenomenology is presently becoming a new source of inspiration for various philosophers of medicine (see chapter VI).

Representatives of all the approaches considered in sections 3 and 4 presuppose that the mental and the physical are primary data, at least at the level of conceptualization. Phenomenology, a *movement* which tried to define primary data in a very different way, is almost disregarded by them. We use the term "movement" for phenomenology to loosely indicate affinities among a large number of philosophers, psychologists and physicians, from Brentano (1838-1917) to Merleau-Ponty (1908-1961) and Van den Berg (1914-), who never formed a "school" as they disagree over most fundamental concepts. Yet they shared a general approach, which we will presently introduce.

The movement has long been confined mainly to the European continent, where it had a marked influence on medicine (cf. chapter II), psychiatry and psychology. A quite different situation has developed in recent years. Phenomenology has lost much of its influence on science, pure and applied, and it hardly affects the philosophy *of science* (articles by Heelan, 1987, and Rouse, 1987, are recent exceptions). But it is still an influential school of philosophy, now especially in the U.S. There it is beginning to function as a new source of inspiration for medicine and psychiatry (see e.g. Engelhardt, 1973; Schwartz and Wiggins, 1985). Indeed a recent issue of the Journal of Medicine and Philosophy (1984, no. 1), which united authors from the U.S., Israel and Western Germany around the theme of "embodiment and rehabilitation", was published in honour of a phenomenological psychiatrist, E.W.M. Straus (1891-1975).

We will not give an exhaustive coverage of the phenomenological movement (Spiegelberg, 1960, gives a good survey). What follows is a personal selection of views defended by authors who may differ in many respects but who share the notion that human existence or "conscious being" is our natural primary datum. We will present these views without criticism, but general comments will be given in section 7. The impact of phenomenology on the philosophy of medicine is discussed in the next chapter.

Phenomenologists do not want to put questions about mind and body at the centre of philosophy. They would argue that one should not be content with concepts like "mind" and "body" before having tried more direct approaches to human being. The experience of changing boundaries between self and non-self opens up such a direct approach.

We are mostly *immersed* in situations, but at times we are confronted with objects, even as we concentrate on ourselves (our *selves*). Experience tells us that we primarily live as embodied selves rather than selves having a body. I wash *myself*, until in the act of washing I come upon an unexpected lump and at once part of my *self* is objectified. I *have* a breast with a suspected tumour. Or I walk in the street, stimulated by a beautiful spring to conduct *myself* as a youth, until someone's look makes me aware of my age. Then I suddenly *have* a bald head and an overweight body. At such moments, the spontaneous unity of being gives way to the distinction of subject and object, of mind and body, by an accident, by somebody's looking at me, by self-observation or reflection. This is where science *begins*. Science does not know about original unity. The physician's clinical look, inspired by science, causes the patient who brought his ailing self to the office to have a stomach, maybe with an ulcer. And because science pervades our culture, the patient may already have been saying to himself, I don't feel well these days, could it be my stomach?

The primary experiences considered by phenomenologists are not restricted to "embodied being" in a strict sense. Embodiment also involves "being in the world" and "being with and for others". When I walk in a forest in autumn, my embodiment may extend to the forest, because I am one with the trees around me and the leaves under my feet. Here too an accident, meeting somebody, or pure reflection, may disrupt the original unity and establish an "internal" consciousness which is distinct from the "external" world which contains my own body.

Changes in the other direction, from objectified to unreflected being, are of course equally common. In daily life, there is a continual switching from one mode of being to another. Our relation to others is an important factor in the process of switching. Consider the following quotation from Van den Berg (1955, p. 55).

As an individual, I always feel more or less strange to my body. I see that it has a certain shape, not particularly desired nor undesired either, and that it, equally without my will being concerned, possesses very definite characteristics of behaviour. I have to accept the shape and the qualities much as I should have to accept the uncertainties of a climate, I go on regarding them with a certain suspicion. Until somebody else teaches me that I may also *be* this body that I *have*, may exactly be it as it is. In the caress the accidentality of the body is removed, there takes place a justification of my body; the caress wipes out the distance between myself and my body, there occurs an adhesion of myself and my body, I begin to inhabit this body, I am invited *to be* this body.

The description we gave so far, however straightforward, cannot be the whole story. It does not explain the priority which phenomenologists give to one mode of being over another. Why could one not regard pre-reflexive being-in-the-world as the lesser mode, "primary" only in the sense that it is a primitive phase? Its "immediacy" might be an illusion which precedes one's cognition of the real self and the real non-self. Perhaps the reflexive cogito, with its subject-object dichotomy, represents mankind's adult approach to reality rather than "alienation".

What mode of being merits the status of "primacy" and why? The question has haunted phenomenology in all its attempts to base a world-view on the primacy of an unfragmented human existence.

5.2. Justifying the primacy of human existence: J.H. van den Berg

The question referred to above may be answered in terms of ethical (or esthetical) preferences or in terms of (metaphysical) essences. The first alternative is the common one. Precisely this seems to have made phenomenology attractive (from the 1930's until the 1950's) for many European intellectuals. The idea of a prereflexive and prescientific being-in-relations (being one with a life-world, with fellow-men, with one's fate and finiteness), appealed to many who felt lost and frustrated in a world dominated by science and technology, without much room for any "search for meaning". This frustration did inspire Husserl's monumental analysis of "the crisis of European Science" and his remedial, transcendental phenomenology (1935, but fully edited only in 1953; cf. Husserl, 1976). Husserl, however, had metaphysical aspirations, hence the "transcendental" nature of his phenomenology.

Other authors emphasized morals rather than metaphysics in reacting to the "crisis of European Science", or, more generally, the crisis of European culture. Most radical in this respect is the Dutch psychiatrist J.H. van den Berg in a series of books on "the shifting existence of western man". Only a few of Van den Berg's books are available in English, but a general synopsis of his line of thinking (with references to translations up to 1968) is given by Jacobs (1968, 1969), and there is a more recent succinct summary by Van den Berg (1978) himself.

Van den Berg writes a cultural history of western man, in which developments in the arts, religion, science and other aspects of culture, are interpreted as changes in the real world. This requires some explanation. Most phenomenologists would describe such developments in terms of *experienced* reality. They would abstain from judgements about the true (objective) nature of the world. In early writings, Van den Berg still shared this position.

The relationship of man and world is so profound, that it is an error to separate them. If we do, then man ceases to be man and the world to be the world. The world is no conglomeration of mere objects to be described in the language of physical science. The world is our home, our habitat, the materialization of our subjectivity (1955, p. 32).

Several years later, however, in a Dutch version (1964) of the same book, he inserted the following passage. "The fact that a thing bears the same name in different situations, does not guarantee that it has the same invariable identity in these situations" (1964, p. 39, translations ours). In the same edition, he discusses psychiatric symptoms commonly described as hallucination or delusion. In either case psychiatry wrongly stipulates that the patient is distorting reality. Thus it confounds two realities, the reality of who is healthy and that of who is ill (1964, p. 102). Van den Berg concludes his argument with the thesis that "phenomenology describes reality" (p. 116).

In fact, Van den Berg in an earlier book (1956) already replaced traditional phenomenology by his own "metabletica" or doctrine of the changeability of man and reality (for reference, see Jacobs, 1968, 1969). We will illustrate what it means by introducing some ideas which Van den Berg presented in later books.

In principle, the structure of reality is twofold. There is the structure which is experienced directly and that which is created with science. The first structure reveals the changing nature of man and world, the second is literally an artefact created by instruments. When one says that some woman is ill because microscopical examination of a blood-film confirms a diagnosis of malaria, one is talking about artefacts. She *herself* is ill and she may get well again, *that* is her lived reality (cf. the first structure). New blood-films may confirm her recovery, but they neither have the positive colour of health nor the negative one of disease. They are second structure *derivatives*, a neuter kind of reality. "The two structures are convincingly related, but their difference is equally convincing. The microscope with all its paraphernalia is interposed between the two structures. It dissociates the object from its original dimensions and robs it of its meaning" (1965, p. 14, translation ours).

In his phenomenological period, Van den Berg, as he notes in 1968, still subscribed to an idealistic philosophy which hypothesizes things "in themselves", admittedly beyond our cognitive reach. This view is now replaced by an endeavour to describe *reality*. He no longer speaks of "phenomena" but of concrete reality in its first structure, the changing reality of daily life. Science, with its principles of invariance, would not dismiss the idea of an atom bomb explosion in the Middle Ages as *incoherent*. The event only could not have happened then for lack of knowledge and technology. But was the first nuclear explosion, on July 1945, really determined by knowledge and technique? Or had matter itself become ready for the event? "Maybe both the theory and the equipment are part of an effort induced and rendered feasible by matter, a changed and modernized matter. Any transference of theory and equipment to the Middle Ages would have to involve a change of Middle Age matter, and this indeed could have made the atom bomb feasible..." (1968, p. 106, translation ours).

Van den Berg regards the "first structure" of reality as primordial because it confers *meaning*. Second structure reality is an artefact which cheats us out of the meaning of life. Medical science is a striking example. It continually adds "health" and longevity to human life without bothering about the fundamentals of human existence. Van den Berg is here obviously taking an ethical standpoint.

In a two volume study of the history of human embodiment Van den Berg (1959, 1961) shows how, in the course of various centuries, the body came to take the shape of an artefact (cf. 14th century anatomy, Harvey's experimental physiology, 19th century discoveries of röntgen photography and of spinal reflexes). Throughout this development, man's natural (first structure) body was gradually emptied of soul and sense, and matter was substituted for meaning. Descartes' dualism was a formal recognition of the changes suffered by the body, long before röntgen rays and reflexes completed the process. "The patellar reflex rendered the body un-inhabitable for the soul. I cannot see how we could inhabit a bundle of nerves and muscles, nor can I see how we could be housed in a body of which the outside is similar to the inside..." (1961, p. 276, translation ours). This expression of Van den Berg's preference for existential being over scientific reality almost sounds like moral indignation. It resembles

Merleau-Ponty's famous dictum: "La science manipule les choses et renonce à les habiter" or (in a free translation), science manipulates reality and renders it uninhabitable (1961, p. 193).

5.3. The ambiguity of human existence: Merleau-Ponty

Merleau-Ponty, like Van den Berg, puts prescientific reality at centre stage. But his approach, unlike Van den Berg's is primarily ontological. In his "Phénoménologie de la Perception" (1945), the relation between reality in itself and the reality which we *live* in perception and actions is an ever returning question. His posthumously edited "Le visible et l'invisible" (1964) shows how he remained occupied with this question until the end of his life. A working note of March 1961, two months before his death, contains the key to his ontological position: "We must describe *the visible* as something realized through man ... and Nature as the other side of man (as flesh - certainly not as "matter")" (1964, p. 328, translation ours).

Farber (1967) has argued that this is a subjectivistic, idealistic position. We do not agree. True, man's subjective mode of being is central in Merleau-Ponty's philosophy. But Merleau-Ponty emphatically founds it on an objective reality. This calls for some clarification.

We experience reality only through continuously changing perceptions which are conditioned by our senses, our perspective, our preoccupations and a host of other variables. Yet all the while we trust that we are seeing things as they are, that we are not enclosed in ourselves but are reaching out to the world, a natural world which we share with others. That is, we assume that there is a world with an objective reality "in itself", a world which includes my body and the bodies of others. But we cannot *at the same time* recognize others as being "for themselves". Objectifying thought ("la pensée objective") recognizes only two alternative modes of being, the *in itself* of spatial objects and the *for itself* of consciousness. When I look at others as beings in themselves, I cannot see them as beings for themselves. "If I constitute the world, I cannot think of another consciousness, for this would similarly have to constitute the world. And, at least from the other world perspective, I would not be constituting anything" (1945, p. 402, translation ours).

This is an unpalatable conclusion, so our premisses must be wrong. Merleau-Ponty submits that it is the strict dichotomy of object and subject, of being in itself and being for itself, which is at fault. The human body should be recognized as a third mode of being. It is neither an object immersed in the material world nor a consciousness positing the world, but a structure enabling the appearance of both world and consciousness. We have a world because we live it, through a body-for-ourselves. This body is a primordial phenomenon through which we are in touch with the objects in the natural world and the consciousness of others in the social world.

This ontology is intrinsically and deliberately ambiguous. It defines the human body first as the primordial source of human existence. Then it posits the anatomical and physiological body as created by scientific reduction, derived from the primordial body by impoverishment. Yet this objectified body has a status of its own in our perceptions. Only because we drag it with us as a material object are we able to take notice of other objects as well (1945, p. 403). The same ambiguity crops up as we feel that things are not "given" to our perception but are "lived" by us. Yet as we "live" them, we make them take the shape of an objective reality because our body is geared ("a un montage") to the world. Thus the human body is a primordial phenomenon for itself, but it is as well attached to a natural world in itself (1945, p. 377). The following quotations (translation ours) further illustrate this fundamental ambiguity.

Things and the world exist only as lived by me or by subjects like me, because they are the succession ("l'enchaînement") of our perspectives. Yet they transcend all perspectives because this succession is temporal and unfinished (1945, p. 385). [The continuous flux of time guarantees the openness and inexhaustibility of being.]

Whether we speak about my body, the natural world, the past, birth or death, the question always is: how is it possible for me to be open to phenomena which extend beyond myself and which yet exist only in so far as I take them up and live them... (1945, p. 417). [Merleau-Ponty then argues that being for oneself, even as it constitutes our reality, destructs subjectivity.]

We posit ourselves, as natural man, in ourselves *and* in the things, in ourselves *and* in the other, so that ultimately, by a kind of crossing-over (chiasma), we become the others and we become the world (1964, p. 212).

The ambiguity, as we said, is deliberate. Let us explain why Merleau-Ponty gives it such a central role. As a phenomenologist, he of course rejects the dichotomy of mind as consciousness *versus* body as matter. Human existence, as he views it, always involves a dynamic balance of the in itself and the for itself ("l'ordre de l'en soi et l'ordre du pour soi"). Sometimes the two modes of being are indistinguishable. But there are also episodes with a breakthrough of corporeality, or a dominance of consciousness in a personal act (cf. our descriptions in section 5.1). "In concreto, man is not a psychism joined to an organism, but a va-et-vient of existence which now lets itself be body, then again sets itself to personal acts" (1945, p. 104). Having thus posited man at an intersection of objectivity and subjectivity, between a natural world of objects and organisms and a cultural world of persons and their life histories, phenomenology rendered his existence ambiguous. We know the world through our body, but as we do that the knowing body is at once reduced to objective corporeality in the world we know.

"How can I understand simultaneously that things are correlative with my knowing body and that they negate it?" (1945, p. 375). Realism has to be rejected since it would destroy perception: "In order to perceive things, we have to live them" (1945, p. 376). But idealism is also unacceptable because things constituted by our perceptions could not be distinct from us. Thus perception would loose its subjectivity and things their transcendence and opacity. "Living a thing is neither coinciding with it nor decomposing it in thought ("ni la penser de part en part") (ibid.). Ontology therefore needs to be ambiguous. What, then, is primordial? Neither the world in itself nor the phenomenon of bodily existence will do. The relation between them is what counts. Nature and man are two sides of the coin of being.

5.4. Moving man in science and philosophy: Buytendijk

The phenomenon of thought easily induces the awareness of a mind-body dichotomy. Yet thinking is intimately linked to speaking, and speaking to facial expression and gestural behaviour. And gestures shade into other activities of the body, including autonomous reflex activity.

Expressive aspects of body movements may indeed give us a glimpse of the mind incarnate, the mind which is not yet detached from the body by thoughts and words. Hence the age-old interest in physiognomy and body structure as marks of human personality, exemplified by Da Porta's (1540-1615) *De humana physiognomia*, Carus' (1789-1869) *Symbolik der menschlichen Gestalt*, and Kretschmer's (1888-1964) *Körperbau und Charakter*. However, empirical studies of *body language* are fairly recent. We will briefly review the origins of these studies in the 1940's and their subsequent fate. The emphasis will be on research by F.J.J. Buytendijk in the phenomenological tradition.

During the second world war, Buytendijk, the Dutch physiologist and phenomenological psychologist, found leisure for a project which had always appealed to him in earlier work on human and animal behaviour. He started writing a book on human posture and movement in which he tried to transcend physiology and psychology. The book was published in 1948 under the title *Algemene theorie der menselijke houding en beweging* (General theory of human posture and movement; translation of quotations given below are ours).

In the preface and the first chapter, Buytendijk discusses a basic methodological problem he had run into. On the one hand, he was convinced of the need to transcend the dichotomy of physiology and psychology in order to uncover fundamental meaningful relations between the living body and its environment (1948, p. 14). On the other hand, he wanted to accomodate empirical data of ordinary science since they are needed to understand "the aspect of blind necessity in human life" (p. 14).

This problem stays with the author throughout the book. The first chapter summarizes his goal as follows. "A theory of human behaviour can only be developed by an anthropologically founded psychology, starting from the essence of human existence and its concrete social relations" (p. 21). Human behaviour, however, is tied to movements of the body which obey deterministic laws of nature. Buytendijk accordingly distinguishes "functional" interpretations, which are in principle future-directed, and descriptions of "enabling mechanisms" which yield *post factum* explanations. The functional view accounts for movements in terms of intentions to achieve something. It is concerned with meaningful interactions between the self and its surroundings. This approach yields "possible knowledge", knowledge of a general kind which cannot be verified. Thus one needs the experimental approaches of physiology and psychology as a supplement.

The book contains a wealth of physiological data concerning reflex behaviour, the integration of neuromuscular mechanisms serving complex locomotor functions, factors shaping the perceived "body image", etc. But the author argues that such data cannot yield satisfactory explanations of action and movement in real life. In neurophysiology, movements are the outcome of a complex interplay of sensory stimuli and motor impulses regulated by various inhibiting and facilitating mechanisms. In real life, body and environment form an indivisible system in which movements are modulated by past experiences, present conditions and intended effects.

Buytendijk feels affinities with representatives of German *Lebens-philosophie* like Scheler and Klages, and also with Von Weizsäcker (1940), who introduced the concept of *Gestaltkreis*. The latter author describes the body and its surroundings as a system of "circular", functional (meaningful) relations. Body movements remodel such relations, they are not mere spatial displacements. As Buytendijk tries to elaborate on these "holistic" claims, his accounts remain globally interpretative. Accounts of specific sequences of postures and movements necessarily revert to reports on experiments with isolated muscles, spinal animals, to clinical observations on cerebral lesions, etc.

Buytendijk's discussion of posture and movement in relation to age and sex exemplifies the global nature of his phenomenological observations.

If we want to visualize the flow of movement of an adult, the straight line is most representative. ... Motion has a distinct starting point (and starting posture) and goes straight to its goal and end. ... A child's movement, however, is chaotic, like an irregular sinuous line with no sharply marked beginning or end (p. 463).

The male gait shows sharp divisions between clearly separate parts that come to a close. ... the male gait can be represented by repeated tapping; such acoustic imitation is not possible when a woman walks" (p. 502).

Female movement can be characterized as uniform, fluent (p. 503).

Such descriptions contrast with physiological details, e.g. those concerning flexion/extension reflexes involved in walking (p. 183-204).

Buytendijk characteristically ends his book by observing that all laws, rules and norms tell us little about the human body. Facial expression, posture and behaviour reflect our entire life history and here "... the diversity of our personal fate rules supreme" (p. 566). His eloquent final words show that what the author really wants to convey is beyond verbal explication.

Subsequent developments. The war which set Buytendijk writing in the seclusion of occupied Holland also occasioned a pioneering study in the U.S. which was to be the first of many empirical studies on human *body language*. D. Efron set out to test the claim of Nazi anthropologists, foreshadowed by many earlier racist authors, that gestures and, more generally, psychosomatic typology, are determined by racial inheritance. For this purpose Efron gathered a wealth of observations on the gestural behaviour (under natural conditions) of population groups in New York city. Using films, sketches by an artist, and descriptions by "naive observers", he compared traditional Jews (mainly first generation immigrants), traditional Italians, assimilated Jews (the offspring of immigrants) and assimilated Italians. His material convincingly shows that culture (living conditions and social contacts) has a dominant influence on gestural behaviour.

Efron's work is important because it represents the first application of an analytic methodology to a subject which had long been treated in a purely speculative way (Kretschmer's typology is a famous example). His book, "Gesture and environment" (1941) was subsequently quoted as a classic in many studies on subjects like "body motion communication", "non-verbal behaviour", "kinesics", etc. We will not review these studies, which range from linguistic analyses (Birdwhistell, 1970) and sociological ones (Scheflen and Scheflen, 1972) to case studies of non-verbal *versus* verbal communication (Ragsdale and Fry Silvia, 1982). They jointly represent an empirical, pragmatic approach which leaves underlying philosophical problems unsolved.

Buytendijk did not want to divorce the study of the lived body from philosophy, and he paid the price for it. Throughout his research he remained saddled with an untractable mind-body problem. For those who, later on, followed Efron's lead, the price was apparently too high. They concentrated on empirical science with practical applications (e.g. in psychotherapy; see Scheflen, 1973), and this apparently precludes philosophical accounts of the phenomena studied. Conversely, those authors who have recently invoked phenomena of body language in support of a mind-body philosophy, revert to a form of intuitive reasoning that is hardly compatible with the spirit of empirical science (cf. Young, 1984). The case of body language once more demonstrates that empirical data do not yield answers to fundamental philosophical questions.

6. THINGS WHICH DON'T FIT

In chapter III, we have contrasted regular medicine with alternative medicine. The emphasis was on homeopathy, an "alternative" which is relatively close to regular medicine. But we also considered, rather briefly, a more outlandish form of alternative medicine, psychic healing. We concluded that one must beware of underrating the limitations of western culture. Other cultures may well come to correct the world-view of the

West. Psychic healing is typically a phenomenon (if it is one) which scientific views of the mental can hardly even place. It is but one of the phenomena which defenders of alternative *science* would point to in criticizing ordinary science.

Throughout this chapter, we have stayed within the modern western tradition. Nowhere in this tradition did we find a satisfactory solution of "the" mind-body problem. Would traditions outside our culture have a better way of dealing with the mental vis-a-vis the physical? We would not like to exclude that possibility. But we want to add that one must be very critical in moving outside the culture one knows. Many of those who are not satisfied with the prevailing scientific world-view turn to the East. They often make a caricature of the culture they try to assimilate, without even trying to test the new theories they embrace.

The confrontation of western and eastern thought which is now taking place in the West is complex. Altered states of consciousness are an important theme which is approached in many different ways. In some quarters sweeping statements about a new world-view abound (cf. the New Age movement). But there is also much rigorous "ordinary" research which simply aims at describing in physiological terms what happens, say, in meditation. And there are those who modestly try to put their own field on a new base (e.g. various clinical psychologists who want to give their discipline a basis in Buddhism; cf. Claxton, 1986).

We do not want to analyse all these developments, but we will present one example, out-of-body states which many people claim to have experienced near death, which shows how "foreign" phenomena might come to influence western views of the mind-body problem. Similar experiences have always received attention in the "occult" tradition, and in the research tradition of parapsychology (cf. experiments concerning putative contacts with the deceased). This alone may explain why regular science and philosophy have disregarded them. The academic community does not like occultism at all, and that should not come as a surprise. There is much humbug on the market, and fraud is common. But let us be open-minded. The possibility remains that some "occult" things are real.

Parapsychology often shares the fate of occultism. Many scholars tend to drop it like a hot potato. Wrongly so. The pressures brought to bear on parapsychology have forced it, long ago, to keep the emphasis on rigorous experimentation besides anecdotal evidence. Recent analyses suggest that one had better take paranormal phenomena seriously (see e.g. the review by Eysenck and Sargent, 1982, and the evaluation of anecdotal evidence by Braude, 1986, a philosopher of science). Our example will anyway suggest that the paranormal may be relevant for the subject of this chapter.

From the 1970's onwards, there has been a shift in research on near death experiences (NDE's for short) to the medical domain (see articles in Greyson and Flynn, 1984, and Gabbard and Twemlow, 1984, and the books by Moody, 1975, and Sabom, 1982). NDE's are characteristic experiences which are often reported by persons who came close to death, after being unconscious or even clinically "dead". The psychiatrist Moody (1975), who investigated many cases, noticed the following elements. First of all, there is a feeling of peace and well-being. One then finds oneself out of the body, passes through darkness (often a "tunnel"), encounters a presence ("a being of light"). There is also a confrontation with a different, beautiful world in which spirits of beloved are recognized. (This, of course, is but a succinct summary.) NDE's typically have such stages, but the later ones need not always be present. The overall pattern is surprisingly uniform.

NDE's have been interpreted in many different ways, hallucination and a touch of after-life being obvious termini of the explanatory spectrum. The picture that emerged during the past few years suggests that we had better allow for a spectrum that extends beyond the boundaries of orthodox science. Out-of-body experiences are perhaps the most telling issue. Sabom (1982), a cardiologist who investigated many cases of NDE with painstaking rigour, concluded that such experiences cannot now be explained in any conventional way. Moody, 1975, and others earlier came to the same conclusion, but their studies were less rigorous.

Some patients, after having had a NDE, were able to accurately report events that took place when they were in coma or unconscious during an operation. The events are described as witnessed from positions that the patient reached "after leaving his body" (a corner of the operation room, an adjoining room, etc.). The investigations showed that the following conditions were apparently satisfied. (i) Reports about many specific details (e.g. conversations among staff; technical equipment) are accurate. (ii) The events described took place when the patient was unconscious (often in deep coma or clinically "dead"). (iii) Events as described could not have been observed from the position of the patient's body. (iv) The patient had no access (e.g. through communication with other persons) to the relevant information before the report was delivered.

Such phenomena are now registered in the context of conventional research. The professionals involved usually demand that brain physiology should ultimately explain the NDE data. As yet, it is unclear what form such explanations could take. So it is tempting to speculate a bit. Would it be conceivable that such phenomena will help to confirm substance dualism, the underdog among current philosophical theories?

Speculate indeed. Our analyses in sections 3 and 4 suggest that empirical research is not the proper tool to decide major philosophical issues. From the perspective of science, relations between the mental and the physical remain an enigma. Some scientists see a clear difference between the two realms. Others regard the distinction as a confused manner of speaking. We would think that one can only resolve the issue by personal choice.

7. DISCUSSION

Dissection of phenomena is an important ingredient of western science. Ideally, the bits and pieces it uncovers are to be covered again by theories with a unifying force. Good theories, in order to bring unity, must be relatively simple.

Today, the accumulation of bits and pieces, of many different kinds, is so rapid that science tends to become unbalanced. One would like to have integrative theories which account for diverse data to retain coherence. But diversity and coherence are hard to combine. Many scientists do *preach* integration through a multidisciplinary approach, but they can hardly practice what they preach. Integrations themselves tend to come in bits and pieces.

We do not think that approaches which characterize science will be able to deal with "the" mind-body problem. Anglo-Saxon philosophy, with its current emphasis on science, is not very helpful either. Its analogon of scientific dissection, *analysis*, predominates to such an extent that it can hardly avoid fragmentation. Philosophy in the hands of Anglo-Saxon scholars tends to forget that generality is a healthy antidote for specialization in science.

Perhaps there is a deeper reason for the failure of science and allied philosophy to really come to grips with relations between the mental and the physical. Thomas Nagel (1986), for example, has argued that objectivity, the hall-mark of science, precludes an account of subjectivity.

Subjectivity as a *perspective* on the world cannot belong to the world *at the same time*. Objective descriptions, therefore, are always incomplete.

There are attempts to redress the balance in science which go by the name of *holism*. Such attempts mostly result in *programs* for the development of theories rather than substantive theories (see section 5.4.; chapter VI).

Phenomenological philosophy represents a different attempt to undo the fragmentation wrought by the intellectual attitude which dominates our culture. It assumes that one may arrive at a coherent view of the world if one concentrates on unfragmented human existence rather than two fields of experience, the mental and the physical, as a primary datum. Phenomenologists regard the mind-body problem as a result of a methodological or a cultural choice to make reality dichotomous.

Phenomenology has been immensely popular (in Europe) a few decades ago. Thus it almost seemed to effect a wholesale reconstruction of psychology. Why has its power dwindled so much? Perhaps the *language* used by phenomenologists has become a stumbling-block. Those who favour Anglo-Saxon approaches will indeed not easily capture what phenomenologists (and, more generally, representatives of philosophies which are typical of continental Europe) are trying to say. Why should the formulations used by them so often have the appearance of obscurity? We think that there is an excuse. Phenomenologists have tried to say what can hardly be put into words. That would also explain the paradoxes which often characterize their philosophy (cf. Merleau-Ponty's "ambiguity"). For us, this is not a negative characterization. Human life, perhaps, is fundamentally paradoxical.

There is yet another explanation for the diminished popularity of phenomenology. Modern scientific psychology tends to make the aims of phenomenology problematic It has shown convincingly that we must be very critical of "primary" *experience*, the ultimate basis of phenomenology (see section 4; chapter III, section 5.3).

Primary experience is permeated by theories, of folk science, popularized science, and general philosophy. Such theories may help one to understand the meaning of life, but they don't mix well with empirical science.

This does not mean that "real science" is the better bet after all. We would rather say that each approach is viable *in its proper context*. In various chapters, we have argued that the search for a single integrative approach is futile. One will have to make choices, different ones in different contexts. In our culture, one choice seems to be basic. The choice between looking at life in wonder, searching for its meaning, and analysing it scientifically, in search of control.

CHAPTER VI. MIND AND BODY IN MEDICINE

1. INTRODUCTION

This chapter will give concrete form to the previous one. There we discussed the mind-body problem against the background of "pure" science
and philosophy. Here we will do the same in the context of medicine.

Is it necessary for medicine to deal with the mind-body problem at
all? One could argue that medicine is only concerned with biological aspects
of health and disease, so that it need not encounter the problem within its
domain. We would not accept this argument, but we admit that it is not easy
to refute it since the definition of medicine's domain is to some extent a
matter of *decision*. However, the argument would not have much force
even if it would be sound. Health and disease do have non-biological,
"mental" aspects at least in the minimal sense of aspects not covered by
current biology. Now the study of such aspects could be relegated to
psychology. Then medicine indeed will not encounter the mind-body problem *within* its own domain. But it will have to face it in reflections *about* its
domain and in confrontations with other domains. Anyway, analyses in
chapters II-IV showed that relations between the mental and the physical
are important for medicine, broadly conceived.

In the present chapter we will first concentrate on a restricted area,
psychiatry and psychosomatic medicine, where one would expect to find
articulated views of mind-body matters. General philosophical approaches
to medicine, e.g. those inspired by phenomenology, are subsequently considered.

Psychosomatic medicine deliberately tries to elaborate a synthesis of
biology and psychology within medicine. Psychiatry is different. It has
schools such as "biological", "psychodynamic" and "behavioural" psychiatry, which are so far apart that a synthesis seems hardly feasible.

Biological psychiatry mostly concentrates on brain physiology (e.g.,
pharmacology of neurotransmitters in the brain) and genetics. Major
mental disturbances such as schizophrenia and other psychoses get special
attention because they seem to offer the best opportunities for a biological

approach. Materialism sometimes seems to be accepted as a basic philosophy, but it is not defended in an explicit way. Psychodynamic psychiatry is less well-defined. It is essentially a modern derivative of Freudian psychoanalysis. At a theoretical level it is often combined with American humanism and/or European existentialism and phenomenology; so philosophy plays an explicit role. The influence of somatic factors on mental phenomena is not denied, but it gets less attention than the inverse relation. Psychodynamics, unlike biological psychiatry, covers a very large area since it tends to function as a general philosophy of life. Behavioural psychiatry is not very explicit about philosophy, though it has inherited the materialistic atmosphere of behaviourism in psychology. It emphasizes the role of the environment, and tends to depreciate the physiological approach. Neurotic disturbances rather than major psychoses get most attention.

We will concentrate primarily on biological psychiatry because it is closest to ordinary medicine. Psychodynamics, and its ancestor psychoanalysis, are considered in later sections.

2. THE MENTAL SUPPRESSED: BIOLOGICAL PSYCHIATRY

2.1. Plain biology or a covert philosophy?

Biological psychiatry is based on the conviction that there is a class of mental disorders representing organic pathology. Representatives of this school agree that the elaboration of nosologies and, more generally, the demarcation of health and disease, are anything but easy in psychiatry (see also chapter IV). But the biological nature of severe mental disorders, according to them, is beyond dispute.

The biological view of mental disorders is supported by various kinds of evidence. Firstly, mental disorders are often associated with characteristic structural and functional features of the brain. Especially neurochemical processes (cf. the role of transmitters, hormones) are implicated. Schizophrenia, affective disorders, and other major mental disorders seem to involve impairment of physiological function. There is a lot of disagreement over details, and the physiology involved is complex (for details, see e.g. Andreasen, 1984; various articles in Weller, 1983;

Halbreich, 1987). Secondly, drugs that influence the relevant physiological systems sometimes give marked improvements in psychiatric patients. What we know about their effects seems to fit physiological explanations of mental disorders (for details, see the same texts). Thirdly, some evidence suggests that genetic factors influence mental disorders (e.g. schizophrenia; comments will be given in section 2.3).

Most defenders of biological psychiatry regard neurochemical factors as major causes of mental disorders, though they are willing to admit that the environment will act as a modifier. The emphasis on biology reflects a philosophical commitment which mostly is not justified in an explicit way. The following quotations (which admittedly represent an extreme example) illustrate this.

During consciousness, a subject is not equally aware of all the multitude of sensory perceptions reaching his brain. How does the brain concentrate? ... Concentration presumably involves selectively restricting incoming information, and a possible mechanism could be the control exerted by the reticular nucleus of the thalamus (Craggs and Carr, 1983, pp. 60-61).

In the traditional model, movements are initiated by the precentral 'motor cortex'. ... However, it seems unlikely that purposive movements do in fact originate in the motor cortex. ... ... movement evidently involves other structures. ... The function of the basal ganglia in controlling voluntary movements appears to be largely an inhibitory one ... (idem, p. 64).

The mechanisms by which emotions are generated are difficult to investigate. ... The dopaminergic fibres may carry impulses generating drive (e.g. appetizing odours arriving through olfactory afferent pathways), which have in themselves become secondary reinforcers by association with eating. ... Eating disorders such as anorexia nervosa may perhaps be produced by abnormal drive mechanisms in the hypothalamus ... (idem, pp. 70-71).

A materialistic philosophy is clearly visible in such passages, but one can only guess at underlying arguments.

Will associations between mental disorders and, say, certain kinds of chemical depletion in the brain justify explanations in purely biological terms? Clark (1980), a psychologist, is among those who have defended this position. He suggests that we can give an adequate neurochemical account of depression (but he does not think that we could eliminate psychological vocabulary). As Russell (1984) has shown, however, such accounts leave us with the unanswered question of *how* biological factors generate mental symptoms. Moreover, they are based on mere associations between biological items and mental symptoms, which need not say anything about etiology.

Effects of drugs, likewise, would not give one information about etiology. They would show that "the physical" can affect "the mental". But it does not follow that causal one-way traffic, from physical to mental, should suffice to explain mental disorders. Moreover, one should be very cautious in accepting claims concerning drug effects (cf. comments on placebo research in chapter III, section 5.6; see also Cohen and Cohen, 1986).

Andreasen (1984, pp. 219-220), after reviewing the evidence, defends a materialistic philosophy as follows.

Like leprosy, mental illness strikes at something that is specifically and peculiarly human. Leprosy rots away the face and hands. Mental illness affects the personality, the way people relate to each other, and the way they think and talk. It *does* seem to affect the spirit, the psyche, the soul. Confusion about the causes of mental illness has arisen because many people tend to assume there must be a dichotomy between the mind and the body, and that a disease of the mind must be different from a disease of the body. People cannot help what happens to their bodies, but they *can* control their minds or spirits.

As the foregoing chapters of this book have shown, the mind and body are in fact inseparable. The word *mind* refers to those functions of the body that reside in the brain. When we talk, think, feel, or dream, each of these mental functions is due to electrical impulses passing through the complicated and highly specialized electrical circuits that make up the human brain. The messages passed along these circuits are transmitted and modulated primarily through chemical processes. Mental illnesses are due to disruptions in the normal flow of messages through this circuitry ... .

The evidence reviewed by Andreasen does suggest that mind and body are "inseparable" in some sense of the term. But even tight connections between the mental and the physical need not lead to the philosophy that she subsequently introduces (cf. "The word mind refers ..."). In biological psychiatry, materialism is often *presupposed* without any defense. We think that philosophical presuppositions underlying theories in psychiatry may have a great impact on research and on therapies. Thus materialism will foster a biochemical interpretation of mental disorders, and drug therapies. This may easily generate self-fulfilling prophecies. If one concentrates on biochemical aspects one will get biochemical results!

2.2. The conflation of dichotomies

If philosophical presuppositions of biological psychiatry would be made explicit, the problems we discussed in chapter V would surface again. We will not reconsider them because that would not lead to any new results. Instead we will follow a different strategy. In chapter V our search was for a solution of the mind-body problem. Here we will not try to come to grips with any philosophical problem. We will criticize common trends in biological psychiatry by a pragmatic methodological analysis. The demarcation of disciplines is our main theme.

Representatives of biological psychiatry would not defend the thesis that biology is *totally* sufficient as a basis for psychiatry. At the very least one will need some psychology and social science as a supplement. The following passage is characteristic.

Many schizophrenic patients improve when given any one of numerous drugs which have a dopamine-receptor blocking action, suggesting that the underlying dysfunction is physiological ... . Depression, too, may be alleviated by any of a large group of drugs with a common action, in this case potentiation of noradrenergic transmission, and a fairly straightforward physiological disturbance again seems likely ... . Mania may perhaps result from a converse disturbance of the same system. Of the remaining mental disorders very little is known of the underlying mechanisms so far. It is of course true that 'psychological factors' (past and present personal experience of life events) play a large part in precipitating and maintaining psychiatric disorders: this is true to some extent even in such 'organic' illnesses as epilepsy and dementia. However, in almost every case an underlying genetic susceptibility has been demonstrated ... and such a predisposition must reside either in the properties of individual brain cells or their connections, or both (Craggs and Carr, 1983, p. 51).

Craggs and Carr obviously think that the mechanism underlying psychiatric disorders is a biological one. Notice what form biology takes in their description. The emphasis is on physiological processes, specifically neurophysiological ones, which are held to be genetically determined. But psychological factors also play a role in etiology. Experience of life events is mentiond as a specific example. We presume that "life events" stands for items such as death of spouse, job loss, etc., which commonly feature in life event research. Psychological factors in etiology are thus associated with the psychosocial environment.

Our impression is that Craggs and Carr have a tendency to associate biological aspects of mental disorders with internal causes or predispositions, whereas psychological aspects are linked with environmental factors with a causative role.

This framework for analysing psychiatric disorders is very common. Mirsky and Duncan (1986), who put biological psychiatry in a broader context, formulate it more explicitly in an extensive review of schizophrenia.

The etiology of schizophrenia is accepted by all to be multifactorial; however, whereas most schizophrenologists would agree that a particular combination of biological predisposition and environmental circumstance is necessary for a schizophrenic disorder to develop, there is disagreement as to the relative weighting of biological as opposed to environmental factors. ...   ... recent studies reviewed here provide evidence from controlled investigations that the external family-community milieu interacts with the internal neuropsychological-neurobiological milieu to produce psychiatric disorder in vulnerable persons (p. 292).

The authors seem to identify "biological predisposition" with "internal-neuropsychological-neurobiological milieu" and "environmental circum-stance" with "external family-community milieu". This represents a com-mon conflation of two dichotomies which are fundamentally different, the biological *versus* the psychosocial and the internal *versus* the external. The conflation betrays a tendency to disregard the ecological dimension of biology and to neglect the possibility that physical and biotic aspects of the environment could play a role in the etiology of mental disorders (the popular book by Andreasen, 1984, represents another example).

We have already shown in various chapters (e.g. chapter II, section 4.3; chapter III, section 2.1) that erroneous views concerning the en-vironment are common in medicine and the philosophy of medicine. Parenthetically, medicine is not the only area where the distinctions we mentioned are amalgamated. The phenomenon is also common in dis-cussions about sociobiology (for a detailed analysis, see Voorzanger, 1987 a and b).

Mirsky and Duncan continue with a survey of three areas they consider relevant, neurological and neuroanatomical studies, cognitive psychophysiological studies, and studies of familial and socioenvironmental factors. Studies in the first area have focussed on many different items, all of them potentially involved in etiology (disturbances in regional cerebral blood flow; alterations in ventricular size and/or cortical atrophy; ab-normalities in glucose metabolism; viral brain infections; disturbed hemi-spheric function or interhemispheric relations; delayed nervous system maturation or deficient neurointegration; pre- and perinatal disturbances). The data combine to suggest that brain abnormalities induced by genetic factors are the "biological basis" of schizophrenia.

Cognitive psychophysiology is the second area involved in schizo-phrenia research. One would expect that *this* discipline will provide one with links between the mental and the physical, but Mirsky and Duncan's review shows that the prospects are dim. The psychophysiological approach of schizophrenia concentrates on stimulus-response relationships, and on information processing. Neurophysiologically defined responses play a crucial role. The approach is meant to give us a better understanding of attentional impairment in schizophrenia. And so it may. But the mental in this context is viewed through a near-physiological window. One hardly meets personal experience of life events.

Mirsky and Duncan put the main emphasis on the "biological basis" of schizophrenia. But they admit that familial and socioenvironmental factors, which constitute the third area of research they mention, tend to modify the effect of biological predispositions. Thus their view seems to imply that one basic dichotomy moulds the entire field of research. On one side biology, concerned with "internal" factors only, explains mental illness as a physical phenomenon. Psychological aspects of the mental are put at the other side where they are associated with an environmental approach. The mental should protest by disappearing. No wonder that "biology" gets so prominent a place.

Various authors who contributed to the general review of psychiatry edited by Goldman (1984) reason along similar lines. We will give three examples. Goldman and Forman (1984) argue that diagnosis in psychiatry is essentially a biomedical affair. But diagnosis will not suffice as it characterizes disorders in general terms. Idiosyncracies of patients are equally important, so diagnosis must be supplemented with "individual psychosocial formulation". At the background of Goldman and Forman's view there is the assumption that environmental factors are psychosocial (cf. the conflation of dichotomies we have already described). Moreover, a new dichotomy, generality *versus* specificity, is merged with the other ones. Thus an important premiss is surreptitiously introduced. General statements about health and disease are *ipso facto* biological, psychiosocial factors cannot be covered by general statements (cf. "*individual* psychosocial formulation")! Goldman and Forman do not explicitly defend this philosophical view, they just introduce it *via* one basic distinction.

Schwartz and Africa (1984), in a review of schizophrenic disorders, introduce a classification of factors that play a role in etiology and pathogenesis with two major categories, biological factors and psychosocial factors. Biological factors are subdivided in terms of anatomy, physiology and biochemistry. Again no place is left for biological factors in the ecological sense. Reus (1984), who discusses affective disorders, also presents a classification of this kind.

Similar classifications are implicitly used in many articles published in journals specifically dealing with mental disorders (e.g. Goldstein, 1987, and sources mentioned by him).

Our survey, however telling, is one-sided. In psychiatry there is also research, apparently unknown to many psychiatrists, which includes a biological approach to the environment. For example, major psychiatric disorders have been explained in terms of "nutritional hypotheses" (see e.g. Abouh-Saleh and Coppen, 1986). It is to be hoped that this kind of work will get more attention.

2.3. Genetic puzzles

It is not easy to assess the impact of genetic factors in man. Basic tools in human genetics are the study of family history, case studies focussing on monozygotic and dizygotic twins, and adoption studies. Biological analysis (biochemistry, physiology, molecular genetics) will naturally facilitate the interpretation of differences in phenotypic characters.

Some of the major mental disorders do seem to involve genetic factors. Schizophrenia is the best example. Twin studies and adoption studies suggest that the environment can hardly be responsible for all the *differences* between schizophrenics and non-schizophrenics (for details see e.g. various articles in Weller, 1983). This is perhaps the best argument in favour of a biological approach of mental disorders. So it deserves special scrutiny. Let us forget about the array of dichotomies discussed in section 2.2 and concentrate on one of them, genetic determination *versus* environmental determination.

To begin with, the emphasis on differences needs explanation (see also chapter II, section 4.3; chapter IV, section 2.2). A statement saying that some feature of an organism is genetically determined, or that it is environmentally determined, is hardly informative. *All* features are "determined" by genetic factors *and* environmental factors. Differences in some feature between organisms, on the other hand, may be purely genetic or purely environmental, or anything in between. Statements about genetic or environmental determination, therefore, must always be set against the background of *particular* comparisons. For this reason a thesis like "schizophrenia is genetically determined", without qualification, is very misleading. It should be taken to mean that observed differences between schizophrenics and non-schizophrenics *to some extent* reflect genetic differences *in the population studied.*

If concepts like "genetic determination" are not interpreted in the correct way, relations between disciplines are easily distorted (Oyama, 1985). Many authors, for example, have misconstrued the position of sociobiology vis-a-vis social science for lack of an adequate understanding of genetics (for details, see Van der Steen and Voorzanger, 1984; Voorzanger, 1987 a and b). Sociobiology is then portrayed as a science that regards human behaviour as genetically determined. Social science, on the contrary, supposedly defends environmental determination (or cultural determination). Such a dichotomy is nonsensical.

Researchers concerned with the genetics of mental disorders have managed to avoid such pitfalls. They do concentrate on genetic *differences.* Moreover, they are aware of methodological pitfalls that hamper the estimation of genetic and environmental contributions to some observed difference. Leading investigators are therefore cautious in formulating conclusions about genetic causes of mental disorders. But they tend to agree that *some* mental disorders (notably schizophrenia) are to *some* extent genetically determined. This, however, is not a very impressive result. It definitely does not justify a materialistic attitude which gives genetic *aspects* of mental disorders an over-important role (cf. our remark on self-fulfilling prophecies at the end of section 2.1).

One could object that our analysis must be wrong because various somatically defined diseases are genetically determined in some stronger way. And the same is true, e.g., of mental retardation linked to abnormal X-chromosomes. Indeed there is such a strong form of determination. But one should be careful in stating what it means. In this case as well, differences and comparisons are involved. "Strong genetic determination" of some feature F prevails when people with a particular genetic make-up would have F *in many environments*, whereas people without this make-up would not have F *in many environments*. Knowledge of biology suggests that there are quite a few cases of strong genetic determination. In these cases, genetic factors are rightly considered important.

If one wants to evaluate research on the genetics of mental disorders, one will need to be aware of the methodological pitfalls we discussed. One example will suffice to show how important this is.

Baron (1986 a and b) has given an excellent survey of research on the genetics of schizophrenia. The emphasis in his review is on shortcomings of current approaches (e.g. problematic statistics), and on problems which result from the heterogeneous nature of the disorder. We would like to concentrate on the following passage (Baron, 1986 a, p. 1053).

Finally, a few words on hypothesis testing. As different genetic models may give statistically acceptable fit to the same set of data, our understanding of genetic mechanisms should be contingent on excluding, rather than including, hypotheses, that is, confidence in the relevance of a given hypothesis for a given data set can be enhanced only if alternative hypotheses can be rejected. For this reason, testing several different genetic models is preferable to a test of a single hypothesis.

What Baron says is to the point, and his phrasing is clear and accurate. But one should read very carefully. The expression "for a given data set", which many authors would have omitted, is revealing once one realizes what it implies. Conclusions about contributions of genetics and the environment are always *relative* to populations one has studied! Thus it is conceivable that genetic factors "play an important role" in schizophrenia in western cultures whereas they are almost irrelevant elsewhere. Or, that the role of genetic factors in some population changes as people change their diets. Parenthetically, cross-cultural comparisons of mental disorders will not be easy (see e.g. Kleinman 1980; Rack, 1982; Kleinman and Good, 1985; Westermeyer, 1985; Jablensky, 1987).

What conclusions can be drawn? Mental disorders, like all other features of organisms, should have genetic aspects. As yet, more substantive claims would hardly be justified.

2.4. Conclusion

In the first part of the previous chapter we argued that biology alone is a poor basis for attempts to solve the mind-body problem. That conclusion, of course, will also apply to the setting of medicine. *Biological* psychiatry cannot provide a solution. Here, however, the situation is worse. Representatives of the profession often do not seem to realize that they have an unsolved problem at their hands. They apparently presuppose that a solution is available in some other field of science or philosophy. Now, one could argue that psychiatrists need not bother in their work about abstract philosophical problems. Theirs is the concrete world of patients and therapies. We think that such an attitude would be misguided. *Our survey suggests that implicit philosophy (often materialism) may well be an organizing force in biological psychiatry's approach to mental disorders, a force which even determines what problems are investigated in research.*

And an implicit erroneous view of biology adds to existing one-sidedness. Needless to say, there will be consequences for psychiatric practice as well. The danger is that patients will be viewed as organisms, not persons.

3. WILL PSYCHOLOGY HELP?

3.1. Science or morality?

After having noted the failure of a biologicized philosophy in chapter V, we turned to the science of psychology for help, in vain as it turned out. Here we will mount another attempt.

Within psychology, *clinical* psychology is the analogon of medicine. It often deals with disorders which psychiatrists would place in the domain of medicine. The boundaries between clinical psychology and psychiatry are indeed vague. We will use the term "clinical psychology" in a wide sense which covers any theory allegedly supporting some form of *psycho*-therapy.

In clinical psychology, the connection between theory and therapy is much more intimate than it is in medicine. Theories of medicine can be tested *via* therapies based on them, but there are "independent" tests as well. By contrast, independent tests are hard to realize in clinical psychology. Moreover, it is extremely difficult to evaluate the efficacy of psycho-therapies. The number of therapies is staggering, and most of them seem to be effective in some way. This suggests that effects are not very "specific", so that they need not confirm underlying theories. It has been argued that therapies based on well-confirmed theoretical principles are the most effective ones (see e.g. Malan, 1976). However, opinions on efficacy and underlying mechanisms must be considered with caution since the currently available methodology of psychotherapy research is notoriously tricky (for details and references see e.g. Garfield, 1984; Klein and Rabkin, 1984; Wojciechowski, 1984; various articles in White, Tursky and Schwartz, 1985; Van Dijck, 1986; Karasu, 1986).

There is yet another reason to doubt whether clinical psychology has adequate (scientific) theories. Many have argued that current theories (and therapies based on them) have a *moral* or even religious character. According to Dyer (1986) psychoanalysis and psychodynamics presuppose a moral concept of *character*. Neu (1977, pp. 112-124) had earlier argued that Levi-Strauss' characterization of psychoanalysis as an analogon of shamanism remains unchallenged. An analysis by Wallach and Wallach (1983) appears to show that clinical psychology is infected with an undefended cult of selfishness. And there are various forms of psychology (e.g. the humanistic psychology defended by Maslow, 1954, 1962) that almost take the form of a religion (for a critical analysis see Bregman, 1982).

True, there are problems with values in medicine as well. In the *philosophy* of medicine, at least, many have defended the thesis that medicine is value-laden so that it cannot contain ordinary science. As we have argued in chapters II and IV, this thesis is too strong. "Normativists" have not recognized theories of medicine for what they are, and they have failed to distinguish various roles of values. One can speak about values in a descriptive way without implicating any *normative* view! The situation in clinical psychology is different. Here normative matters appear to play an essential role. We do not mean to say that clinical psychology is no good, but only that much of it does not have the force of ordinary *science* for lack of empirical theory.

The issues mentioned above are often disregarded in clinical psychology itself (see e.g. reviews in Watts, 1985). However, Diekstra, a Dutch clinical psychologist, has explicitly addressed them. He does not consider his discipline as a science (see Diekstra and Dijkhuis, 1985). According to him, psychotherapies based on it differ greatly from medical therapies. We had better put them outside the scope of science. How is the peculiar position of clinical psychology to be explained? Diekstra has an interesting suggestion (op. cit., p. 34). Psychology became a science almost overnight when Wundt started his experimental work in 1879. Before, it belonged to philosophy. Thus it missed an important stage of development which other sciences passed through after they separated from philosophy: the stage of field observation. The best-confirmed theories of psychology are based on experiments in the laboratory which need not reproduce significant aspects of real-life situations. No wonder that clinical psychology can tell us little about everyday affairs. It has failed to acquaint itself with them.

By way of an example, we will briefly introduce one species of psychotherapy, existential psychodynamics, and compare it with biological psychiatry.

3.2. *Existential commitment*

Biological psychiatry scrutinizes the physical, it hardly gives the mental a status of its own. Psychodynamics has moved in the opposite direction. We will briefly consider one variant, existential psychodynamics, as represented by Yalom (1980).

Existential psychodynamics is a synthesis of ideas derived from Freudian psychoanalysis and European existentialism and phenomenology. It also has affiliations with American humanistic psychology.

Freud's dynamic model of mental functioning is by and large accepted by Yalom:

The psychodynamics of an individual ... include the various unconscious and conscious forces, motives and fears that operate within him or her. The dynamic psychotherapies are therapies based upon this dynamic model of mental functioning.

So far, so good. Existential therapy, as I shall describe it, fits comfortably in the category of dynamic therapies. But what if we ask, which forces (and fears and motives) are in conflict? What is the *content* of this internal conscious and unconscious struggle? It is at this juncture that dynamic existential therapy parts company from the other dynamic therapies. Existential therapy is based on a radically different view of the specific forces, motives and fears that interact in the individual (Yalom, 1980, p. 6).

Yalom focusses on four ultimate concerns, death, freedom, isolation, and meaninglessness. "The individual's confrontation with each of these facts of life constitutes the content of the existential dynamic conflict" (p. 8). These are concerns of *man* which are only aggravated in psychiatric patients. Psychopathology therefore cannot be understood unless it is placed in the context of a fundamental philosophy of existence.

Existential dynamics are not wedded to a developmental model. There is no compelling reason to assume that "fundamental" (that is, important, basic) and "first" (that is, chronologically first) are identical concepts. To explore deeply from an existential perspective does not mean that one explores the past; rather, it means that one brushes away everyday concerns and thinks deeply about one's existential situation (pp. 10-11).

The various existential analysts agreed on one fundamental procedural point: the analyst must approach the patient phenomenologically; that is, he or she must enter the patient's experiential world and listen to the phenomena of that world without the presuppositions that distort understanding (p. 17).

This is where European philosophy comes in. The basic distinction of subject and object is rejected. So is the methodology of orthodox science.

For example, the empirical research method requires that the investigator study a complex organism by breaking it down into its component parts, each simple enough to permit empirical investigation. Yet this fundamental principle negates a basic existential principle. ... Furthermore, the empirical approach never helps one to learn the *meaning* of ... psychic structure to the person who possesses it. Meaning can never be obtained from a study of component parts, because meaning is never caused; it is created by a person who is supraordinate to all his parts (p. 22).

According to Yalom, mental disorders represent an ineffective mode of coping with anxiety. Thus he argues that death anxiety plays an important role in schizophrenia.

Existential psychodynamics has inherited many of the problems that plague classic psychoanalysis (for reference see section 5.2). Moreover, it has been criticized on the ground that existentialism and psychodynamics are not really compatible (Hanly, 1985). Such comments are important, but we want to keep them at the margin. Our main concern is that biological psychiatry and psychodynamics (especially the existential variant) are far apart along so many dimensions, that it is hardly conceivable that their representatives can communicate at all. Indeed they don't.

Existential psychiatry, as opposed to biological psychiatry, depreciates etiology.

Existential psychiatry rejects the methodology of common science, which biological psychiatry accepts.

Existential psychiatry puts inner experience at the centre of theories, biological psychiatry keeps it at the margin.

Existential psychiatry comes with an explicit philosophy, biological psychiatry shuns explicit philosophy.

We would not accept existential psychodynamics as a good alternative for the biological approach (which has obvious shortcomings). Its moral commitments would not be acceptable for many patients (cf. Chapter II), and its rejection of any common methodology turns possible grounds for accepting it into quicksand. However, we welcome its emphasis on the fundamental concerns of man. They should not be left out in any attempt to understand mental disorders.

Would it be possible to give existential psychodynamics a scientific basis? It is not easy to answer this question. One may call existential psychodynamics an alternative science, and the regular/alternative dichotomy is complicated (see Chapter III).

Biological psychiatry has an unduly narrow view of inner experience, existential psychodynamics gives it an overprominent place. Biological psychiatry has a biased methodology, existential psychodynamics hardly has any methodology. Neither of these extremes offers a convincing solution of the mind-body problem.

4. THE PSYCHOSOMATIC CONNECTION

4.1. Towards integration?

Biology alone is a poor basis for understanding relations between the mental and the physical. So is psychology. That much became clear in chapter V, and the preceding sections confirm it. In chapter V we subsequently analysed attempts to integrate biology and psychology to come to grips with the mind-body problem. Here we will do the same in the context of medicine.

Attempts to give medicine (broadly conceived) a new basis by fusing biology and psychology have taken many different forms. For example, various investigators (Reiser, 1984; Winson, 1985) have tried to develop new theories which encompass neurobiology and psychoanalysis. We will not discuss such theories since they have a restricted scope: the domain of psychology is represented by psychoanalysis only. The "system approach" and "holism" represent less restrictive endeavours aimed at integration; we will consider them in section 5. We have chosen to concentrate, in the present section, on psychosomatic medicine. In our view this discipline covers the most sensible attempts to develop an integrative approach because its researchers try to *implement* abstract frameworks for integration in the context of medicine. That constitutes the positive side of the psychosomatic coin. However, there also is a negative side. In concentrating on implementation, researchers have often taken the philosophical framework for granted. We intend to show that philosophical presuppositions of psychosomatic medicine are not as unproblematic as representatives of the discipline would have it.

By way of introduction, the following quotation will suffice to convey the *atmosphere* of research in psychosomatic medicine.

Warburton's ... approach to the physiological stress response indicates that this response is *initiated* by cognitive events in the system, events that by the identification of externally and internally generated stimuli *trigger* the adrenocorticotrophic system to supply energy, and stress steroids. In this conceptualization, psychogenic or cognitive stressors precede physiological and psychopharmacological stressors. Once the system is working under the *influence* of stress hormones, the resulting changes in brain chemistry assume a more dominant role by *facilitating* cognitive processes that may either *reduce* ... or *increase* the flow of corticosteroids (Hamilton, 1982, p. 107; italics ours).

Notice how the mental and the physical are mixed in this passage. Causal relations (cf. italics) are posited without a deeper analysis. One should be aware of the background philosophy which is taken for granted here. In this respect at least, biological psychiatry and psychosomatic medicine are not very different.

Would it be necessary to deal with the mind-body problem in a more explicit way in psychosomatic medicine? Most researchers simply do not ask the question, but there are exceptions. *If* the issue is forced into the open, the answers given by different people are very different. Graham (1979) has argued that the mind-body problem does play a role in psychosomatic medicine. If one tries to get rid of it, it simply will take a different form. The troublesome dichotomy of functional and organic disorders is an example. Weiner (1982) seems to deplore the absence of a solution for the mind-body problem in psychosomatic medicine. Lipowski (1984), however, thinks that the mind-body problem is a nice subject for philosophers only. Psychosomatic medicine had better concentrate on practical and methodological matters. Finally, there are those who think that they have dispensed with the mind-body problem.

And now I shall state what is the essence of my thesis: Any neurally mediated response is behavior. ... Since all neurally mediated responses are behavior, it is not necessary to invoke dualistic notions such as biobehavorism to explain how organismic actions of one sort affect organismic actions of another sort (Engel, 1986, p. 472).

"Any ... response *is* behavior". What does that mean? If Engel wants to introduce a *factual* connection, he will have quite a job to confirm what he is stating. But perhaps he is giving a covert *definition*. If that is true he simply *stipulates* that dualistic notions are superfluous.

In the rest of this chapter we will concentrate on science rather than philosophy because that is common practice in psychosomatic medicine.

Research by Selye (1936, 1950) is one of the sources which have influenced psychosomatic medicine. (For the record, we want to mention two other researchers who have been influential, Alexander, 1950, and Groen, 1957). Selye concentrated on *stress*, which has remained an important subject ever since. He defended a *nonspecificity hypothesis* which says that many different adverse conditions produce the same physiological effect. In later years, (non)specificity became a controversial subject (for examples see Lazarus, 1977; Levenson, 1979; various articles in Field et al., 1985; Paré and Glavin, 1986). Likewise, the concept of stress became problematic (see e.g. various articles in Field et al., 1985, and in Zales, 1985). Some investigators even think that it is useless (e.g. Hinkles, 1977; Engel, 1985).

The whole field of research on psychosomatic medicine or, more specifically, stress cannot be covered here. We will selectively choose examples from the literature to illustrate how the mental and the physical are approached.

4.2. Historical notes on stress

The basis for modern theories of stress was laid by Selye in the thirties. His work in subsequent decades forms the core of stress research inspired by biology (for a survey see Selye, 1983). As a student, Selye became obsessed with the idea that diseases, however different, have much in common.

During this whole period I was obsessed with the thought that there existed specialists in every branch of medicine ... but no one had tried to specialize in the common and hence most important syndrome: that of sickness as such. I wondered why the already well-known methods of exact scientific investigation, of looking for quantitatively measurable biochemical, microscopic, or functional changes, could not be employed to clarify the mechanism of 'the syndrome of just being sick' (pp. 3-4).

Ten years later, Selye again encountered the syndrome in experimental work with rats which was to bring him fame. He observed that many different kinds of noxious stimuli (heat, cold, infection, etc.) produce the same stereotyped response, which he called the general adaptation syndrome (GAS). The GAS has three stages, the alarm reaction (increased heart rate, loss of muscle tone, decreased temperature, changed blood pressure; subsequently increased secretion of corticoid hormones), the stage of resistance (disappearance of initial symptoms; adaptation through compensatory reactions) and the stage of exhaustion (illness and eventual death after prolonged exposure).

Selye came to use the terms "stress", for the response, and "stressor" for any stimulus causing it. Stress, in his words, is "a nonspecific response of the body to any demand" (p. 2). Later research showed that the nonspecific response pattern involves many interconnected processes throughout the body.

Psychology meanwhile took a separate share of the stress field. While Selye emphasized responses, psychology focussed on psychosocial stimuli and on the role of personality characteristics. The biological approach initially overshadowed psychological research, but a good balance was reached after Lazarus (1966) published *Psychological stress and the coping process.* Biology and psychology now mix freely, but the marriage is not quite happy. There is much disagreement over facts and also over concepts. We will first concentrate on concepts. Disputes over facts can hardly be settled if we cover them with ambiguous terms.

4.3. The meaning of "stress"

Discussion of definitions seldom plays a prominent role in the life sciences. "Stress" is an exception. It is variously used for stimuli, for responses, and for their interaction, and the merits of alternative definitions are widely discussed. Most authors now reject Selye's definition in terms of responses. They rather want to give the stress concept a heavier load. Let us consider some examples.

Let me start with a psychological, rather than a physiological, definition of stress. External *stressors* are effective to the extent that they are perceived as dangerous or threatening, that is, to the extent that they are cognitively interpreted as inimical. One of the difficulties with the so-called objective definitions of stressors has been their neglect of these cognitive mediators. Mason ..., in a critique of Selye's stres concept, noted that emotional arousal is one of the most ubiquitous reactions in situations that are considered stressful. However, these emotional responses depend on psychological interpretive mechanisms. ... I shall assume ... that the major consequence of such an interpretation or appraisal is the activation of a stress reaction that is psychologically functional. In other words, the consequence must be perceptible. ... Thus, a situation is defined as *stressful* if and when the interpretive cognitive activities of the organism transform the input in such a way that a perceptible internal change results (Mandler, 1982, p. 91).

At first sight, Mandler is concerned with various *empirical* relations. Stressors induce emotional responses. Such responses also depend on cognitive interpretation. The responses are psychologically functional. Therefore, both the stimuli and the responses must be perceptible. These relations, however, are subsequently summarized in a *definition*. So they become unempirical.

We are faced with an awkward dilemma. If we accept Mandler's definition, apparently interesting relations between the mental (cf. cognitive activities) and the physical (cf. stressors) are covered by a *definition.* As a result, an *empirical* investigation of such relations will become impossible. If one would still want to investigate them (and Mandler apparently wants to do that) one would have to reject the definition. But then the notion of stress would again be unclear!

Mandler has confused questions of logic and questions of fact, so he has not really solved any conceptual problem. Analysis of recent literature shows that this confusion is quite common. The examples given below illustrate that.

There is no logical incompatibility that stands in the way of integrating psychological stress theory with Selye's theory. Both approaches are, in fact, complementary. Psychological stress theory outlines the conditions which determine the evocation of stress while Selye's theory describes its form. To portray what is significant in these approaches, the following definition of stress is suggested. ...

Stress is a state which arises from an actual or perceived demand-capability imbalance in the organism's vital adjustment actions and which is partially manifested by a nonspecific respons.

An objective of this new definition is to emphasize the continuity between psychological and physiological theorizing (Mikhail, 1985, p. 37).

Mikhail apparently wants to associate his definition of "stress" with research in psychology. In a survey of psychological stress theory that precedes the definition, he mentions three important aspects of stress which were "identified" in the fifties by psychological research. "1. Individuals differ in their reactivity to stress. ... 2. Stress is determined by the perception of the stressful situation rather than by the situation itself. ... 3. The extent of stress depends partly on the capability of the individual to cope" (p. 35).

The three statements are meant to express empirical relations. However, they also play a role in Mikhail's definition. One is again confronted by a dilemma. If the definition is accepted, the three statements loose the force they are meant to have. If one rejects the definition, one may not understand the statements for lack of a definition of "stress", the central concept.

Leading investigators in the field of stress are aware of conceptual pitfalls. Monat and Lazarus (1985, p. 2), in the introduction of their anthology, characterize the situation as follows.

The reasons investigators have been unable to reach any general agreement on a definition of "stress" are undoubtedly complex but revolve largely around the problems inherent in defining any intricate phenomenon. For example, a response-based definition of stress (e.g., one that looks at increased physiological activity as an indicator of stress) suffers from, among other things, the fact that the same response pattern (such as increased blood pressure or heart rate) may arise from entirely different stimulus conditions, for example, from heavy excercise or extreme fright. And, of course, the psychological meanings of these conditions are typically quite different ... . Likewise, stimulus-based definitions are incomplete because any situation may or may not be stressful, depending on characteristics of the individual and the meaning of the situation for him or her.

Monat and Lazarus' central point is that stress is an intricate *phenomenon*. Our diagnosis is different. Researchers continually try to elaborate an intricate stress *concept*. Intricacy is not an intrinsic attribute of any phenomenon. *Current conceptual problems in the stress area simply result from an unproductive methodology*. One had better characterize various kinds of stimulus, and various kinds of response in independent terms; in doing that, one will perhaps have to abandon the comprehensive concept of stress. Otherwise the formulation of sensible empirical statements about "stress" will become impossible.

The use of a comprehensive stress concept could wrongly suggest that science has managed to integrate biological and psychological theory, and that the mental has thereby found a place in medicine. But mere conceptualization is a poor surrogate for integration.

4.4. Nonspecificity, a testable issue?

Hypotheses, like concepts, must satisfy elementary methodological criteria if they are to make sense. Testability is an important criterion (cf. chapter III, section 3). In order to be testable, hypotheses must be clear. Ambiguities in the stress concept will therefore make hypotheses concerning stress problematic. The literature amply illustrates this. For example, endless discussions on the nonspecificity of stress postulated by Selye have not led to consensus. The following quotation from Mikhail (1985, pp. 32-33) illustrates how conceptual problems obscure hypothesis testing.

As noted earlier, Selye attached great theoretical significance to the nonspecificity of stress reactions and disease. This view contrasts sharply with the older specificity position of Pasteur's theory which maintained that each disease could be induced only by its specific causative agent or microbial infection.

The nonspecificity of the stress response, however, has been the subject of some controversy. Mason (1971), for instance, found that the nonspecificity of the pituitary-adrenocortical activity, the stress response, was not as broad as Selye had suggested. The effects of several kinds of nocuous agents - fasting, muscular exercise, heat, hemorrhage and cold exposure - on the secretion level of urinary 17-hydroxycorticosteroid (17-OHCS) of monkeys were tested. According to Selye's theory, it would be predicted that elevation in 17-OHCS should result from all the noxious agents despite their qualitative differences. Contrary to expectation, neither fasting nor heat raised 17-OHCS level. The failure of 17-OHCS to rise, however, was noted only when psychological influences were minimized - that is, when fasting monkeys were isolated from nonfasting ones and when nonnutritive cellulose diet pellet was given to them to provide some bulk within the gastrointestinal tract. But when psychological factors were included as in the condition wherein monkeys were suddenly deprived of their daily food while their neighbors were eating next to them as usual, a marked rise in 17-OHCS was observed.

The inability of the physically harmful stressors, fasting and heat, to raise 17-OHCS could be interpreted by the opponents of the theory as evidence against it. On the other hand, advocates of the theory could argue that what appeared in this experiment to be a test of different kinds of stressors in eliciting the nonspecific response was really no more than a test of different levels of stressor intensity. The lower levels failed to activate the stress response simply because they were weak. When fasting was strengthened by psychological factors, the nonspecific response occurred.

Mason's experiment did not reach an unequivocal conclusion about the nonspecificity of the stress response because it tested a theory which is not formulated in a testable form. Before we can reasonably be asked to test the nonspecificity notion we should at least be able to induce stress with some certainty. To do so, the conditions which constitute a stressor must be stated independently of the stressor's effect. Not any demand is stressful. Demands that evoke stress are only those which tax the organism's capability and which appear to be of value to the individual. The extent of stress depends largely on the individual's evaluation of the consequences of unfulfilled demands.

It is necessary to point out that the study of stress requires that the researcher or clinician be aware of its psychological determinants. The ambiguity of this aspect of stress in Selye's formulation impairs the design of sensitive experiments and the attainment of reliable findings.

Mikhail apparently thinks that Selye's definition of stress in terms of responses leads to problems with testability because the hypothesis at issue (the nonspecificity hypothesis) can be saved by ad hoc moves if there is negative evidence. His solution, which he does not work out in detail, is formulated in rather ambiguous terms (see the last part of the quotation). "Conditions which constitute a stressor must be stated independently of the stressor's effect." What does that mean? On one interpretation, Mikhail is saying here that "stressor" must be *defined* in a particular way. "Be stated" would then have definitional force. But on a different reading, Mikhail is referring to *factual* connections. "Demands that evoke stress *are* only those ... ." Again it is unclear whether the intended connection is definitional or factual (see also section 4.3).

We want to analyse the situation in different terms. Selye's central thesis says that many different adverse conditions (nocuous stimuli) cause the same response pattern (GAS, the general adaptation syndrome). This justifies the use of the term "nonspecificity". How should one formulate a nonspecificity hypothesis? Nobody in his right mind would suppose that any stimulus whatsoever, at any intensity, will provoke the pattern. Alternatively, a nonspecificity hypothesis could be formulated as follows. "All stressors (nocuous stimuli) result in the GAS (in stress)". But this will not do either. Selye defined "stressor" in terms of "stress", so this variant of the hypothesis is a tautology. One could also try to give an exhaustive specification of conditions that provoke stress; cf. "Conditions C provoke stress", in which C stands for many different items. The principle of parsimony will not allow one to regard this as a fruitful hypothesis. What is left is the vague but informative statement that *many* different conditions produce stress (GAS). If this is what the nonspecificity hypothesis amounts to, it is simply well-confirmed. Selye himself at times suggests that he has a stronger thesis in mind, but this only indicates that some reconstruction of his views is necessary.

How should further research be conducted? Selye concentrated on one particular approach: spelling out the details of the GAS, and embedding it in physiological theory. That is a good thing. One can also try to make an inventory of conditions that provoke the response, so that the ecological context becomes clear. For this purpose one will need to test more *specific* hypotheses with the form "factor X at intensity I produces element E of GAS in species S". Mason's work showed that this approach leads to problems because there are interactions between effects of different factors (cf. the factors presence or absence of food, and presence or absence of neighbours). So more sophisticated hypotheses must be elaborated. Whatever line is taken in subsequent work, one will need independent terms for various stimuli, and for various responses. Mikhail (and many others with him) tried to remedy the vagueness of "nonspecificity" by replacing Selye's reasonably clear stress concept with a much more untractable notion (cf. section 4.3, second quotation). *That* move will lead to problems with testability.

The place of psychology in Mikhail's argument is also revealing. As suggested above, Mason's experimental results could be discussed in plain biological terms. Why should the observed effects of *neighbours* make us introduce psychological factors? Is the psychological approach irrelevant when we consider effects of starvation in isolated animals? In short, when should we pay attention to the mental? Mikhail just introduces physiological and psychological factors in an intuitive fashion without bothering about their relations. He rightly supposes that some factors which are often *called* psychological interact with other factors which are often *called* physiological, but he does not pause to comment on the labels. This is a common phenomenon in stress research. The nature of relations between the mental and the physical is not really addressed.

4.5. Methodological interlude

If the methodological pitfalls we came across would be avoided, our problems would by no means be over. Suppose one investigates the correlation between *social support* and responses to adverse conditions of a particular kind. Support could somehow be measured in terms of interactions between individuals. Would the absence of correlation imply that support may not be an important factor? Not at all. As Wethington and Kessler (1986) have argued, *received* support and *perceived* support can be very different. So perhaps one has not chosen the most relevant variable. It will not be easy to define adequate variables which are not interdependent. There are many methodological problems of this kind. We will not survey them since recent surveys of methodological *minutiae* are already available (Blaney, 1985; Alloway and Bebbington, 1987). Here we will concentrate on a few more general issues.

As concepts and hypotheses in research on stress are made more precise, the chances are that natural history will take the place of *general* theories. Would it be possible then to fuse biology and psychology so that one gets integrative theories? That is a moot point. Perhaps one will not end up with any *theories* at all!

We have already shown, in section 4.4, that an elaboration of Selye's work should lead to the formulation of rather complicated hypotheses which cannot be very general. That suggests that research on stress will move from general theory to natural history as it becomes more sophisticated. Now investigations based on Selye's work represent but one tradition in the field of stress. An overview of the whole field would only strengthen the case for natural history. For the record, we will briefly consider two other traditions.

Alexander (1950), unlike Selye, postulated *etiological specificity* for various disorders. He argued that certain emotional conflicts afflict certain internal organs. For example, inhibited rage seems to affect the cardio-vascular system. At first sight, the approaches of Selye and Alexander seem to be incompatible. Responses are specific or they are unspecific, they cannot be both. However, Alexander's specificity hypothesis is as vague as

Selye's nonspecificity hypothesis. The two hypotheses may turn out to be compatible if they are developed in more detail.

In another line of research (originally psychosomatic medicine, now mostly psychophysiology), a different kind of specificity called "autonomic response specificity" was postulated by Lacey, Bateman and Van Lehn (1953). They continued research by Malmo and Shagass (1949), who argued that psychosomatic diseases often show "symptom specificity". The point is here that *individuals* show idiosyncratic responses (Alexander was concerned with specificity with respect to *stimuli*). Interest in the subject subsequently waned for lack of convincing evidence. However, the existence of specificity has recently been confirmed in some experimental settings of psychophysiology (Van der Molen and Orlebeke, 1980; Orlebeke et al., 1985). It was shown that subjects react to certain stimuli with a clearly idiosyncratic heart rate response. Psychological features, e.g. a trait which may be called "stress tolerance", seem to affect the response to stimuli.

These results should serve as a reminder. Many psychologists have since long argued that (at least for some purposes) stimulus-response paradigms must be replaced by stimulus-organism-response paradigms. *Of course* features of individuals will affect their responses to stimuli. Thus it is probable that relations between stressor and disease are affected by aspects of personality (Eysenck, 1983).

By and large, recent research suggests that empirical generalizations concerning stress are feasible, but their scope will have to be limited. Numbers of stimuli, responses, and features of organisms which may be introduced into stress research are large indeed. One will have to make choices, and the kind of results one gets (e.g. with respect to "specificity") will depend on the context created by the selection of variables. As research continues, contexts will tend to become narrower. Thus natural history may come to replace general theories. A characteristic example of this trend is discussed in the next section. First we want to discuss a different methodological issue with implications for research on stress.

If one really gives psychology the place it deserves in psychosomatic medicine, it will not be easy to dispense with mentalistic concepts. Is it possible to treat them as one would treat the concepts of natural science? To some extent it is, but one will mostly have to work with "non-natural" scales and scores. Consider the influence of life events on morbidity; an important theme in psychosomatic medicine. In the study of their effects, one will want to have a measure for their emotional impact. Emotions qua inner experience cannot be measured in a direct way. So one will have to be content with an assessment based, e.g., on verbal reports. Such information can be used for the development of a scale with severity values for various kinds of event. Death of a partner, for example, will get a high value. Once there is a scale, it becomes possible to assign values (scores for summed important events) to patient histories, and this will facilitate the detection of relations between life events and illness. Needless to say, we have greatly simplified (cf. the survey of methodological problems in life event research by Kasl, 1983), but that does not affect the point we want to make.

"Non-natural" scales and scores are necessary for scientific psychology. They are less pervasive in natural science. What does that mean? There is a simple answer. The phenomena natural science is interested in can often be described with quantities associated with stronger scales. The (weaker) scales of psychology are contrivances which are needed if one wants to study elusive phenomena (cf. inner experience), or if one wants to get some order out of a wealth of heterogeneous data.

We think that methodological differences between biology and psychology will not make integration easy. Consider again the life event example. One can surely arrive at generalizations which connect life event scores with, say, biochemical data. One could even go further and argue that life events are causally relevant. But what would such a statement mean? One should notice that scores can hardly stand for causally relevant properties, they are but a convenient summary pointing to mental processes which we assume are causal.

Even if it would be possible to elaborate general empirical statements concerning observed relations between mental phenomena as studied by psychology and biochemical processes, methodological problems would still abound. Isolated general statements do not have much value. In order to be significant, scientifically or philosophically, observed relations would have to be embedded in a theory which explains how factors in different domains can show interaction. In other words, one would ultimately need some solution of mind-body puzzles. Thus one has to face the issues discussed in chapter V all over again. Anyway, one should not loose sight of the fact that relations between the mental and the physical as psychosomatic medicine describes them are as yet unexplained.

4.6. The specifics of psychosomatics

The present state of the art in psychosomatic medicine is adequately captured by Stein's presidential address for the American Psychosomatic Society (Stein, 1986). He begins with an appraisal of two major theoretical positions concerning specificity, represented by Selye and Alexander, which we have already mentioned in section 4.5. Seleye came with a nonspecificity theory which says that a great variety of stressors have similar effects on physiology and pathology (see foregoing sections). Alexander (1950), contrariwise, postulated specificity for certain disorders.

According to Stein, there is ample evidence by now showing that neither theory is quite acceptable. His own approach also concentrates on specificity, but he puts it in a different context. Many biological functions are characterized by specificity. The immune response is a clear example. Olfactory recognition among animals, likewise, involves specificity of responses. "It appears that specificity is a primitive but critical function of olfaction and immunity, two of the major integrative and adaptive systems of animals" (p. 6).

Olfaction and immunity played an important role in Stein's own research. He showed that asthma attacks, in patients seen in psychotherapy or psychoanalysis, are often triggered by odours. Close scrutiny uncovered "the specific idiosyncratic meaning of odors which was not revealed by an investigative procedure which merely determined and classified the type of odor that will precipitate an attack" (pp. 7-8). That is, one will have to ask patients about personal experience in order to understand why certain odours affect asthma.

Some methodologically sophisticated work with animals confirmed the thesis that odours play a role in asthma. It appears that attacks, in some respects, are like a conditioned response.

Many other relations between psychosocial factors and somatic disease have been established beyond doubt. Epidemiologists have observed that life events such as death of a partner are associated with increased morbidity and mortality. Until recently, the mechanisms involved were obscure. But according to Stein the situation is rapidly changing. There is now much information on interactions between the immune system and neuroendocrine regulation. Emotions involved, e.g., in depression could affect neurotransmittors, and so result in neuroendocrine dysregulation and alterations of immunity.

This excursion has emphasized that there is no single specific factor involved in the etiology of disease, but rather there is a long chain of relevant processes and interactions. Standing as we do now at the vantage point of decades of achievement, it is possible to begin to see the relationships between CNS mechanisms and cyclic nucleotides as intracellular mediators and modulators of various biologic systems. As we acquire increasing knowledge of specific biologic regulatory mechanisms, the processes underlying disease appear to be more complex while at the same time more understandable. The notion of specificity persists, but there has been a shift from the causal specificity associated with a disorder to specificity within the individual (p. 18).

One should notice that specificity as envisaged by Stein is even more idiosyncratic than the specificity investigated in psychophysiology (see section 4.5). Here the issue is not merely that psychological features (which may be enduring traits) affect stimulus-response relationships. Items like "idiosyncratic meaning of odours" are a function of particular episodes in the history of the subjects involved. Thus they will hamper the formulation of generalizations concerning relations between stimuli, reponses, and features of organisms even if the context is kept appropriately narrow.

The prevalence of "specificity" in Stein's sense will function as a constraint upon the elaboration of general theories. True, there are general statements to the effect that immunity in man, and in other species, shows a particular kind of specificity. And there are general ideas about the way in which specificity develops. But generality fades to the background as one concentrates on the fate of individuals. Similarly, as more "relevant factors" are uncovered (cf. the great variety of neurotransmittors, cell types involved in immunity, and so forth) it will not be easy to formulate statements which are both general and precise. Perhaps one will be able to identify factors which are *often* involved in pathogenesis. But statements about the precise role of factors will always have to be qualified by the admission that effects will depend on the circumstances.

Complexity and specificity do spell trouble for scientific approaches if one wants to maintain common standards of explanation, prediction and management. The present proliferation of factors considered by psychosomatic medicine may be useful because it stimulates the elaboration of new hypotheses and theories. However, as the process continues, the discipline will need a new core. The ideas of Selye, and of Alexander, had this function in the past. As yet, there is no adequate successor.

If one would have a successor, one would not thereby have solved any mind-body problem (cf. section 4.4). Theories of psychosomatic medicine associate mentalistic concepts with concepts of natural science (otherwise one would not be dealing with *psycho*somatic medicine) in order to unravel relations between the mental and the physical. But they ultimately leave the relations unexplained.

Science, within the context of current medicine, does not seem to have a satisfactory theory that could make us understand relations between the mental and the physical. Therefore, let us turn to philosophy, as we did in chapter V.

5. THE LIMITS OF INTEGRATION

5.1. The biopsychosocial solution

Our survey of "integrative" views of the mental and the physical in medicine is of course incomplete. We will supplement our discussion of psychosomatic medicine with an analysis of two popular paradigms of integration, the "general system view" inspired by Von Bertalanffy and "holism". The so-called biopsychosocial model of disease (and of health) is our first example. Those who give it a central place in medicine generally favour a general systems view. Engel (1977, 1981), who is a representative of psychosomatic medicine as well, is one of their most important spokesmen.

As argued before (chapter V, section 3.5), Von Bertalanffy's general systems philosophy is best regarded as a program for the development of integrative scientific theories. The program was sensible, but it has remained a program. We will argue that Engel's version of the systems approach in medicine, likewise, is a program that has still to be implemented. In his 1981 article, Engel introduces his approach as follows.

Nature is a "hierarchically arranged continuum, with its more complex larger units being superordinate to the less complex smaller units" (pp. 103-104). If one begins at the top of the systems hierarchy, one gets: biosphere, society-nation, culture-subculture, community, family, two-person, person (experience & behaviour), nervous system, (...), subatomic particles.

Each level in the hierarchy represents an organized dynamic whole, a system of sufficient persistence and identity to justify being named. (...) Each system ... implies qualities and relationships distinctive for that level of organization and requires unique criteria for study and explanation. In no way can the methods and rules appropriate for the study and understanding of the cell as cell be applied to the study of the person as person or the family as family (pp. 105-106).

Every unit in the systems hierarchy is at the same time a part and a whole. If one is to understand any system, therefore, one must consider more inclusive systems besides components. But investigators are obliged to concentrate on a particular system level. "For the physician that system level is always *person* ..." (p. 106).

Engel subsequently presents a clinical example to show how the biopsychosocial model works. A patient, known by the pseudonym Mr. Glover, was brought to an emergency department with symptoms of myocardial infarction. Beforehand, he had first resisted acknowledging illness, but his employer enabled him to accept hospitalization and patient status. After admission, Mr. Glover accepted the reality of another heart attack (it was his second one). Although he was no longer having any discomfort, the staff effected prompt coronary care. Subsequently there were problems with an arterial puncture, so the house officers left him alone, saying only that they were going for help. As Mr. Glover reported afterwards, this made him loose confidence in the staff. Then he suddenly lost consciousness; the monitor documented ventricular fibrillation. Fortunately, defibrillation was successful and the patient recovered.

Engel extensively documents the case to illustrate his approach.

Even such minimal screening data as Mr. Glover's age, gender, place of residence, marital and family status, occupation, and employment indicate systems characteristics useful for future judgements and decisions. The information that the patient resisted acknowledging illness and had to be persuaded to seek medical attention ... reveals something of this man's psychological style and conflicts. From this alone the systems-oriented physician is alerted to the possibility ... that the course of the illness and the care of the patient may be significantly influenced by processes at the psychological and interpersonal levels of organization. [So one needs an inclusive approach.]

Such an inclusive approach, with consideration of all the levels of organization which might possibly be important for immediate and long term care, may be contrasted with the parsimonious approach of the biomedical model. In that mode the ideal is to find as quickly as possible the simplest explanation, preferably a single disease diagnosis ... . For the reductionist physician a diagnosis of 'acute myocardial infarction' suffices to characterize Mr. Glover's problem and to define the doctor's job (p. 108).

In his description of the case, Engel continually emphasizes the role of various levels of organization. Thus he comments as follows on the episode before the cardiac arrest. "His [Mr. Glover's] account raised doubts that the onset of ventricular fibrillation could be ascribed solely to processes restricted to the injured myocardium alone. Rather it suggested a major role for extracardiac (neurogenic) influences originating in disturbances at the two-person and person levels" (p. 114).

Engel subsequently uses the example to contrast the biopsychosocial model (i.e. the systems approach) with the biomedical model.

The emergency room approach was conventionally and narrowly biomedical. It was predicated on the reductionist premise that the cause of Mr. Glover's problem, and therefore the requirements for his care, could be localized to the myocardial injury. This, plus the high risk attendant upon such injury, justified proceeding with the technical diagnostic and treatment procedures with only passing attention to how Mr. Glover was feeling and reacting. ... [This approach is associated with the model of controlled experimentation, and, in clinical practice, a sequential "ruling out" technique which focuses on one issue at a time.]

A systems approach to helping Mr. Glover would have differed in notable respects. From the outset the decision to provide immediate coronary care would have included consideration of factors other than cardiac status, notably those manifest at the person level. The interview of Mr. Glover would have been conducted in such a manner as to elicit simultaneously information needed to characterize him as a person and to evaluate the status of his cardiovascular system. This could have been readily accomplished by having the patient report symptoms in a life context, noting activities, reactions, feelings, and behavior as symptoms were evolving, as well as the circumstances of his life preceding the onset of symptoms. ... The difficulty with the arterial puncture would have been recognized as a risk for the patient, not just a problem for the doctors. Mr. Glover's failure to complain would have been anticipated as consistent with his personality style ... . Whether such an approach would in fact have averted the cardiac arrest is impossible to know. But certainly sufficient experimental and clinical evidence exists linking psychological impasse and increased risk of lethal arrhythmias ... (pp. 115 and 120).

Engel emphatically calls the biopsychosocial model a scientific model. The biomedical model, likewise, is scientific, but it has degenerated into a dogma. Only the biopsychosocial approach represents adequate science.

We wonder how Engel would define the term "model". He introduces the biopsychosocial model in the form of diagrams which purport to show that there are relations between levels of organization. But nowhere does he state what the relations are. He assigns events to various levels of organization, but he does not formulate general statements which characterize processes at any level (see also Schwartz and Wiggins, 1985). This suggests that Engel's model is an unimplemented program, an outline for a model or a theory. If he would try to develop a theory in greater detail, he would doubtless have to face all the problems we uncovered in our discussion of psychosomatic medicine.

We concur with Engel's criticism of what he calls the biomedical model. One needs more than biomedical theory and biological data as a basis for medical practice. But would anyone deny that? The real issue is, what alternatives or supplements are available? There are no integrative theories of the mental and the physical. Engel suggests that he has one, but his descriptions contradict this. Alternatively, one could work with unconnected or loosely connected theories from various areas of science. Lastly, one could use a biomedical approach and top this up with common sense. Our impression is that Engel is actually following the latter strategy.

5.2. Holistic medicine

"Holism" is a label which is adopted by many of those who want to expose the narrow biological approach which is common in orthodox medicine. Holists come with alternatives which can take various forms. Firstly, there is holism in a rather narrow sense of the term which is almost equivalent to the systems approach. Engel is a holist in this sense. He himself does not use the term in this way. Indeed he says that he rejects holism. Unfortunately the terminology in the literature is confused.

Other varieties of holism, which now flourish in the western world, are a mixed lot. Many of them are concerned, in one way or another, with the mind-body problem. By and large, the approaches they represent are covered by theses discussed elsewhere in this chapter or in the previous one. Indeed it is not easy to evaluate holisms because they cover so much territory! The following succinct caricature may not be quite undeserved. Orthodox medicine, as holists see it, wants to be concerned with biological aspects of health and disease only. Holists want medicine to be concerned with everything.

As holism is gaining power especially in the U.S., opposition is becoming intense. Thus papers in the book edited by Stalker and Glymour (1985) seem to represent a veritable crusade against holism. The book does contain good criticism, but its general approach is rather biased. Kopelman and Moskop (1981) give a more balanced account of the holistic movement in the U.S. We will comment on their analysis to indicate how theses defended by holists are connected with themes we discussed in other chapters. Our analysis will be brief because holism is not specifically concerned with the subject discussed in this chapter. But we had to give holism some place.

The various forms of holism have a common core. Kopelman and Moskop mention five tenets which they regard as characteristic. Firstly, health should be defined positively in terms of well-being, not in the usual negative way. Secondly, individuals ought to be encouraged to take responsibility for their own health and illness. Thirdly, health care providers should have an important role in education. Fourthly, health care delivery systems must deal with behavioural, social and environmental causes of illness. Fifthly, well-being must be promoted primarily with natural or non-invasive techniques.

These common features still allow much variation. The spectrum of holistic views is broad indeed. At one end, a conservative variety only emphasizes psychological measures and lifestyle modification to compensate for limitations of conventional medicine (e.g. Pelletier, 1979). At the other end, there are those who accept any odd alternative treatment modality without critical analysis (e.g. LaPatra, 1978).

What about the role of science? Kopelman and Moskop suggest that the two extremes represent alternatives with respect to critical appraisal within the holistic movement. One either accepts scientific tests of modalities of healing, or one accepts a plethora of treatments in an uncritical way. Kopelman and Moskop are positive only about the first option. We do not quite agree because their description is misleading. It seems to presuppose that a critical attitude is *ipso facto* a scientific attitude and *vice versa*. However, those who defend alternative therapies may do so on the ground that they have a legitimate criticism of the common methodology of science. Kopelman and Moskop seem to think that the methodology of science can be taken for granted. As we have argued in chapter III, that is a problematic assumption.

Kopelman and Moskop's comments on the first and the second tenet of holism are biased in a different way. Consider the first tenet, which identifies health with well-being. The authors point to one consequence of this identification which is especially troublesome.

A great deal of the gross national product (10%) is spent on health ... . If health means well-being, then is anyone who does anything that makes us feel better or instructs us in activities that would do so (sailing, skiing), engaging in healing practices? If so, should they then receive health insurance payments? (...) ... while it is perfectly acceptable to make reformative or stipulative definitions of health in terms of well-being, the consequences of this in terms of funding health care systems and "alternative" health care systems are problematic (p. 223).

We think that this assessment is simplistic. Even if health care is made very inclusive (cf. sailing, skiing), facilities available for patients may be used in a selective way so that costs remain reasonable. Some types of sport training are indeed incorporated in medical programs (e.g. in Holland, Denmark) which are less expensive than the American ones. The argument in the present form is anyhow defective. It presupposes that the current investment in health care is sensible. But one must not forget that expensive high technology need not really promote health (see chapter II). Kopelman and Moskop do not analyse the impact of alternative health care systems, so their argument begs the question.

The second tenet says that persons ought to be encouraged to take responsibility for their own health and illness. The authors comment as follows.

Least controversial are the views that beliefs, attitudes, dispositions, habits, life-style and environment have *some* effect on health and illness. More extreme and counter-intuitive is the view that persons are entirely or almost entirely responsible for their health and illness (pp. 223-224).

The extreme view cavalierly assigns blame for illness and seems to re-establish the ancient notion that misery, sickness and death are linked with moral failure and sin (p. 224).

Implicit moral and religious assumptions also appear to underlie the emphasis on individual responsibility in at least some of the holistic health literature. ... In contrast, orthodox medicine does not appear to require the adoption of an implicit moral or quasi-religious commitment to health as an absolute value (p. 225),

We agree with the letter of these statements. But their spirit is something else. Consider the italicized "some" in the quotation from pp. 223-224. It could suggest that the effects involved are minor ones. But suppose that there are reasons for reading "enormous" instead. Where would that leave us? True, it does not follow that medicine will have to accept the far-reaching morality which some holists seem to adopt. But the thesis that their kind of moralistic medicine is inadequate does not imply that it is possible for medicine to be non-moralistic. Kopelman and Moskop do not actually come with this implication, but their text suggests that they would accept it. One anyway needs a *comparison* of various kinds of commitment. We agree that some forms of moralism in the holistic health movement are unacceptable. But we do not want to assume without argument that orthodox medicine is the better alternative. As we have shown in chapters II-IV, the issues are not that simple.

Holism, in short, is concerned with almost all the subjects discussed in this book. This diffuseness, presumably, is what tends to make discussions about it confused.

6. THE PHILOSOPHICAL TURN

6.1. Playing with phenomenology

Phenomenological philosophy emphatically aims at the dissolution of the mind-body dichotomy. It has been very influential, but by now its power has dwindled. At its culminating-point, it visibly influenced medicine in Europe. We have already discussed the ensuing paradigm, anthropological medicine, in chapter II.

The program of anthropological medicine was fascinating, but it was never really implemented. The last book by Buytendijk (1965), a Dutch representative of anthropological medicine, has the telling title "*Prologomena* to an anthropological physiology". Buytendijk was 78 years old when it was published. He could but plead that others should excecute the program which he had sketched. Von Weizsäcker eventually realized that the world-view which anthropological medicine envisaged could not be articulated. *It has to be lived* (Von Weizsäcker, 1956, pp. 171-178).

After the demise of anthropological medicine, there have been new attempts to give the phenomenological approach a place in medicine. The work of Engelhardt is a case in point. His treatise on the mind-body problem (1973), however, is not meant as a basis for a new *theory*. He primarily wants to emphasize an *attitude* of respect and understanding. Engelhardt describes how our culture has partitioned man over various disciplines. The mind-body dichotomy is one of the results. No place is left for unity of being and unity of meaning in any particular discipline. One has to make choices if one wants to study man in a disciplined way. Medicine has apparently chosen to concentrate on the body. And its approach has been successful within the chosen domain. But one should not overlook the one-sidedness of this approach. There is an increasing temptation, in medical theory and practice, to reduce man to physical parameters and to treat him accordingly.

How could one explain the waning of phenomenology's influence? We have already hinted at various possible causes in the discussion of chapter V. Phenomenology tried to capture experience uncompromised by any artifice of science or, more generally, preconceptions. Philosophers in other schools would argue that such a thing is impossible, but philosophical arguments seldom really settle disputes. However, recent developments in science strengthened the case against phenomenology. Academic psychology showed that *theories*, of academic science or folk science, permeate experience. This makes the phenomenologist's search deeply problematic. Moreover, Husserl's thesis that it is possible to get rid of preconceptions in a methodical way was soon criticized within phenomenology itself. Merleau-Ponty's notion of *ambiguity* in fact points to the impossibility of uncontaminated experience.

In the last few years, the influence of phenomenology in medicine seems to have increased again. We will briefly discuss some examples.

Leder (1984) argues that medicine's Cartesian paradigm must be replaced by a phenomenological "paradigm of embodiment". His approach resembles that of anthropological medicine. Medicine must concentrate on the "objectified" body if it is to effect cures. But it must do much more than that. The body is first and foremost a *lived* body, a "unity of sensori-motor intentionality". If one forgets that, one will get a distorted view of health and disease. Consider the following examples.

A person may express a subliminal grasp of the world as overpowering, the self as inadequate before it, by assuming a stooped body posture. Years later the chronic back problems that result come before an orthopedic surgeon. Another body takes up the classic pose of fight or flight in a stressful office situation. Over time the internal hypermobility leads to gastritis, high blood pressure, perhaps a heart attack. The oversecretion of acid, the constriction of an artery exhibits the expressiveness of bodily movements no less than motions externally manifested (p. 39).

Leder does not deny that a biomedical approach can be satisfactory in the treatment of illnesses with relatively pure "physicalistic origins". But these are "limiting cases of bodily intentionality" (p. 40).

What consequences should the phenomenological paradigm have for medical treatment? Leder argues that many new therapies have appeared, both in conventional and in alternative medicine (e.g. biofeedback, yoga, visualization techniques to combat cancer, primal scream therapy).

Though it is not always recognized, these therapies operate with an implicit concept of the self closer in spirit to that of Straus [a psychiatrist who worked in the phenomenological tradition] and Merleau-Ponty than Descartes. ... Such therapies would by no means supplant the traditional options modern medicine offers. Nor will all prove efficacious.

However, they are too often neglected because our Cartesian heritage does not yield a structure by which they can best be understood. The theory of bodily intentionality helps create a space in which such forms of treatment make eminent sense. If applied to medicine, the paradigm of the lived-body might assist in the explanation and further development of such novel therapeutic techniques (pp. 41-42).

We agree with many of the theses Leder defends. Conventional biomedicine is one-sided. It abstracts from "lived experience", and this leads to a biased view of health, disease and illness. Conventional therapies have their limitations. Perhaps various "alternative" therapies are a good supplement. And so forth.

However, it would not be easy to implement Leder's program. How can one be sure that the stooped person who enters the consulting-room has a "subliminal grasp of the world as overpowering"? How should therapies of alternative medicine be tested? The usual answer given by alternative healers is that he who heals is right. That may make sense against the background of some criterion of success. But alternative healers often do not articulate a criterion. On the other hand, if the answer is that one will have to go at the problem with scientific methods, one will be back at ordinary medicine, or medicine supplemented with ordinary psychology, or perhaps psychosomatic medicine. That will reintroduce many of the problems which the phenomenological approach was designed to avoid. There is no integrative scientific theory which really clarifies relations between the mental and the physical.

Alternatively, one could argue that scientific interpretations will not suffice. The views of phenomenology are needed as a supplement. However, this will not do either. There are no *articulated* phenomenological prescriptions.

There is yet another way to reconstruct Leder's view. Conventional approaches in medicine inspired by science concentrate on particular aspects of health and disease, as they must, so they are one-sided. One needs common sense and intuition to redress the balance. We would agree, but we would not like to use the label "phenomenology" for this view. And we would add that common sense and intuition are fallible.

Carpenter (1986), in a comprehensive survey of schizophrenia research, also defends a phenomenological approach, but in a less emphatic way. He convincingly argues that the purely biological approach of schizophrenia is dangerously one-sided. Psychosocial aspects are as important as biological ones. Divergent perspectives must somehow be integrated in clinical practice. In that context we need an approach which is associated with the phenomenology attributed to Husserl and Jaspers. Carpenter describes it as follows.

In clinical practice, it is time to return to a basic empirical approach. The irreducible essence of our interest in schizophrenia is the nature of another person's experience. It is in the subjective and inner world of volition, perception, cognition, and affect that schizophrenia is manifest. Empathy is crucial in discovering the subjective life of another person, and to engage in this process with the psychotic requires skill, intuition, training, and perseverance. Jaspers recognized that knowledge of this inner world was central to diagnosis, prognosis, and treatment, and he borrowed phenomenology from Husserl's philosophy and introduced it into medicine as a scientific method ... . Phenomenology defines the goals and methods used by the clinician to generate the observations upon which clinical decisions are based. When this base is weakened by narrowly defining psychopathologic attributes, or by distorted observation with premature inference and theoretically determined assumption, our view is truncated and the science of clinical care is compromised.

We agree, but again we think that "phenomenology" in a philosophical sense of the term is not at issue here. "Skill, intuition, training, and perseverance" are indispensable. But could they be covered by a clearly defined phenomenological ("scientific") *method*? We doubt it. Carpenter anyway does not describe such a method.

Other recent attempts to put medicine on a phenomenological basis, at least those known to us, can be criticized in the same way. We will briefly consider one other example. Schwartz and Wiggins (1985, 1986b) argue that medicine needs the human sciences besides natural science. The methods of the two kinds of science are different. Natural science aims at explanation. The human sciences use the method of *understanding*. (This is a thesis of hermeneutics; see section 6.2.) Science, however, is not enough. Both science and humanism are derivatives of the "lifeworld", which can be grasped only by a phenomenological approach.

The lifeworld is the sphere of pre-scientific experience. ... It is the realm of everyday social interaction and practical projects. Here we do not conceive the world through scientific ideas; we rather perceive it through our senses and engage in it through bodily activity. We encounter our fellow humans and communicate with them while we engage in common practical tasks. ... The medium of this communication is the natural language that we inherit from our shared cultural tradition, not the technical and formal language of science (1985, p. 340).

We do not think that pre-scientific experience can have the autonomy vis-a-vis science which the authors postulate.

Schwartz and Wiggins argue that the difference between science and ordinary life "lies primarily in the determination of scientists to base all their beliefs and practices strictly on *evidence*" (p. 345). But in ordinary life there is prescientific knowledge. "This prescientific knowledge of reality is never erased by science and replaced with the latter's "improved" view. Scientific thinking rather draws on this foundation of pregiven meaning in order to reach beyond it" (p. 348). This stance naturally leads to the question of *how* the "drawing on" can be articulated. The authors do not answer this question. They disregard popularized science as a source of common sense in our society. One would also like to know when pre-scientific knowledge can be trusted. Again there is no answer.

We have much sympathy for Schwartz and Wiggins' approach. But it seems that it will reintroduce all the problems concerning relations between science and common sense (and intuition) which we considered before.

Margulies (1984) would also want to give phenomenology an important place in medicine (at least psychiatry), but his approach is different. He argues that phenomenology (Husserl) *and* psychoanalysis (Freud) resemble *art* rather than science.

Both Freud and Husserl devised methods that *should* create a dialectic tension at the very heart of their hard-won systems of knowledge. In the spirit of truth and investigatory honesty, discord was ensured. Paradoxically, the more one learns, the more one must be able to put aside. The very design of these methods, if adhered to, prepares the observer for the opportunity of perceptual novelty. The methods are inherently creative and artistic (p. 1030).

On this view, the "methods" of phenomenology must not be put alongside methods of science. With this we agree. (Psychoanalysis is different; we will come to that.) The common suggestion that the phenomenological approach is an *alternative* for approaches that centre on "normal" science is misleading. Likewise, an integration of the two kinds of method seems hardly feasible. They call for different attitudes which we cannot have *at the same time* though we may need both.

We have a critical comment as well. *Many* different methods may inhibit cognitive enslavement (Margulies' term) and so stimulate perceptual novelty and creativity. The choice of a method will have to be a personal one. We would not like to mount a *general* defense of any particular method.

6.2. Afterthoughts on psychoanalysis and hermeneutics

The survey we gave in sections 4 and 5.1 is one-sided in two respects. Firstly, the views we mentioned in section 4 combine to suggest that clinical psychology and psychotherapy have few points of contact with science. However, one should not forget that psychotherapy was originally put by Freud in the context of rigorous science. Secondly, our discussion of phenomenology is incomplete since we did not deal with a philosophical school which has put phenomenology in a new perspective, *hermeneutics* (for a survey, see Thompson, 1981). Hermeneutics is interesting in the present context since it has contested the thesis that Freud's psychoanalysis contains any science.

"Hermeneutics" is often broadly defined, so that it comprises Habermas' critical philosophy and Ricoeur's hermeneutics *sensu stricto.* It draws on various philosophical schools which have flourished in continental Europe, existentialism, phenomenology, Dilthey's idealism (cf. the distinction of *Geisteswissenschaften* and *Naturwissenschaften),* and to some extent marxism (Habermas). This sets it apart from Anglo-Saxon philosophy, especially philosophy of science. The gap between the two traditions is further emphasized by different attitudes towards science. Humanities and arts (Ricoeur) and sociology (Habermas) are important sources of inspiration for critical philosophy, whereas natural science has often been a paradigm of intellectual rigour for philosophers of science in the Anglo-Saxon tradition.

In medicine and the philosophy of medicine the impact of hermeneutics is on the increase, specifically where social aspects of medicine are considered (for a survey see Wulff, Pederson and Rosenberg, 1986). We only want to mention reactions of hermeneuticians on psycho-analysis which bear on our present subject, the mind-body problem. For further information one should read Grünbaum's excellent analysis of these reactions (Grünbaum, 1984 and 1986b; see also the peer reviews which follow the 1986 article).

Hermeneuticians regard the therapeutic process in psychoanalysis as a process of self-emancipation which cannot be captured by natural science. Both Habermas (1971) and Ricoeur (1981) have argued that Freud failed to see this due to "scientistic self-misunderstanding". Freud, so they argue, wrongly forged a link between his speculative *metapsychology* (which has the flavour of science) and *clinical* concepts which are outside the scope of science altogether. In the clinical setting one can only reach *understanding*, not scientific *explanation* because scientific explanation is non-historical. The laws of science simply do not apply to what happens in psychoanalytic therapy, because causal laws are overcome in the process of self-emancipation (Habermas). Moreover, scientific theories on the therapeutic process could never be tested because the client is the ultimate arbiter of success or failure. There can be no intersubjective validity (Habermas again). Ricoeur's arguments follow a similar course. He concentrates on *verbal* productions of clinical interactions, and emphasizes *understanding* outside the domain of scientific testability.

Grünbaum's comments are crushing, rightly so. For one thing, Freud himself emphatically dissociated his metapsychology from his clinical theory. The metapsychology, which is formulated in terms of abstract notions such as *id, ego, superego*, is a speculative superstructure of the clinical theory, which centres on repression. The clinical theory is the most important part of Freud's thinking. He explicitly says that it can be tested without recourse to metapsychology, and he gave a clear account of the methodology he used in actual tests. We think that Grünbaum has convincingly shown that hermeneuticians have gravely distorted Freud's writings. Their analysis could at best apply to some aspects of recent varieties of psychoanalysis, e.g. the "heuristic" approach defended by Peterfreund (1983).

To make things worse, Habermas and Ricoeur have distorted natural science as well; it is simply not true that causal explanations in science are always "non-historical". The same goes for Anglo-Saxon philosophy of science. For example, Habermas makes a caricature of the notion of testability. And Ricoeur has at best succeeded in *making* psychoanalysis "untestable" by first removing the parts which ensure testability. And so on, and so forth.

In the second part of his 1984 book Grünbaum shows that Freud's psychoanalysis is actually testable. According to him, it appears to have been *falsified*. This thesis is more contentious. Many philosophers (e.g. Popper) have argued that the theory cannot be scientific because it is untestable. We will not take sides. As argued in chapter III, testability remains an elusive criterion.

7. CONCLUSIONS

We can but conclude that the comments at the end of chapter V still stand. "The" mind-body problem, which has always haunted western culture, is still with us today. Science and philosophy have not provided a substantive solution. Meanwhile, our analysis of the situation in medicine has uncovered more specific problems which do not receive the attention they deserve in medicine or in philosophy of medicine.

Firstly, biological approaches of mental disorders, however useful, are often one-sided even with respect to the biology they are based on. Psychiatrists at times contrast biology *in the sense of physiology and anatomy* with an environmental approach which is relegated to psychology and social science. The biologist's ecology is thereby defined out of existence. This can hardly lead to a balanced view of the etiology of mental disorders.

Secondly, attempts to integrate psychology and biology within medicine (cf. psychosomatic medicine) are not as succesful as casual inspection of the literature suggests. Basic notions such as "stress" are often given a heavy theoretical load by *definitions* which refer to items from biology *and* psychology. The ensuing "integration" is merely pseudointegration. One cannot amalgamate distinct theories by linguistic tricks.

Mind and body are seemingly further apart in medical theory than in pure science and philosophy. Criticism of the "biomedical model" is on the increase. But philosophers of medicine have a tendency to develop "alternative" views of medicine in a somewhat uncritical way. For example, phenomenological views presented by them are "phenomenological" in a rather diluted sense of the term. They often strike us as common sense embellished with philosophical terminology.

We like to suggest that analyses in the spirit of Anglo-Saxon philosophy, of concrete examples from medicine itself, should get a more important place in the philosophy of medicine (cf. the examples presented in sections 2 and 4). They may uncover limitations of current approaches in medicine which one easily overlooks. True, philosophical *analysis* has clear limitations. It will not result in a coherent world-view. But if one tries to develop a general philosophy without the benefits of analysis, one will easily succumb to uninformed vagueness.

CHAPTER VII. THESES

The theses presented below cover many of the subjects we have discussed. They are meant as a guide for readers who want to concentrate on themes of particular interest in teaching or research. Chapters and sections are indicated after each thesis.

Medical science and medical practice

1. Medical practice is more than applied science. Common sense in various forms is indispensable. *II.2.2; III.5.3; IV.5.*

2. Views which reflect common sense (intuition, experience) in medical practice are often called "phenomenological". The label is harmless but it must not be used to suggest that there is a clearly defined phenomenological method. *II.2.2; VI.6.1.*

3. Common sense is permeated with theory. Specifically, it involves folk psychology which may need to be corrected by academic psychology. *III.5.3; IV.5; V.4.1.*

4. Applied science, in medical practice, has its limitations in view of uniqueness and subjectivity of patients. But it would be a mistake to think that science cannot cover uniqueness and subjectivity in any way. *II.2.2.*

5. There is no convincing integrative theory of mind and body, in medicine or elsewhere. *V, VI.*

6. Biology does not suffice as a basis for mind-body theories. *V.3.*

7. Biological psychiatry is an inadequate basis for understanding mental disorders. *VI.*

8. Psychosomatic medicine has no integrative theory. It easily takes the form of natural history. *VI.4.*

9. There is no adequate concept of stress. *VI.4.*

Health and disease

10. Cure is not the heart of medicine. *II.2.*

11. Ecology, as a discipline of biology, does not get the attention it deserves in medicine. There is a fallacious tendency in medicine, and in the philosophy of medicine, to identify biology with anatomy, physiology and genetics, and to associate the environment with psychology and sociology. *II.4.3; III.2.1; VI.2.2.*

12. Achievements of medicine are but a minor factor in improving health at the population level. Philosophy of medicine had better take this into account. *II.4.*

13. Health and disease are *always* influenced by genetic *and* environmental factors. Therefore, "genetic determination" and "environmental determination" easily become misnomers. *II.4.3; IV.2.2; VI.2.3.*

14. Health as adaptation to the environment is a misnomer. *II.4.3.*

15. Typological approaches of health and disease are misguided. *IV.2.2.*

16. Health and disease have non-biological aspects. But this does not imply that concepts of health and disease, and concepts for diseases, cannot make sense if they are defined in biological terms. *IV.3.2.*

17. Explications of concepts for health and disease are context-dependent. *IV.3.2, IV.3.3; IV.4.2.*

18. The "biopsychosocial" approach of health and disease represents common sense rather than medical science. *VI.5*.

Facts and values

19. Philosophical views of medicine are often ambiguous for lack of an appropriate distinction of facts and values. *II.2.3; IV.4.1*.

20. Medical theories can be empirical even if they deal with values. *II.2.3*.

21. Facts are not sufficient as a basis for decisions concerning values in medicine, but they play a vital role in such decisions. *II.4.4*.

22. Health and disease are values. But that need not imply that the *concepts* of health and disease must be value-laden, or that statements about health and disease have normative force. *IV*.

23. From a logical point of view, "normativism" and "naturalism" may well be compatible. *IV.4.1*.

Miscellaneous subjects

24. Medicine is a poor basis for world-views and general views of man. *II*.

25. Various philosophies of medicine contain idiosyncratic elements of cultural bias. *II*.

26. Alternative medicines must not be rejected *merely* because they do not satisfy methodological principles endorsed by regular medicine. Such principles have visible limitations. Specifically, testability is a problematic criterion. *III*.

27. Outlandish phenomena must not be dismissed *simply* because they cannot be covered by ordinary science. *III.6; V.6.*

28. Adequate tests of medical theories and medical treatments cannot be simple. *III.5.6; IV.5; VI.4.4.*

29. The concept of "holism" in medicine is confusing because it is over-inclusive. *VI.5.2.*

REFERENCES

Abou-Saleh, M.T., and A. Coppen, 1986. The biology of folate in depression: implications of nutritional hypotheses of the psychoses. *Journal of Psychiatrical Research* 20, 91-101.

Agich, G.J., 1983. Disease and value: a rejection of the value-neutrality thesis. *Theoretical Medicine* 4, 27-41.

Alexander, F., 1950. *Psychosomatic Medicine.* Norton, New York.

Alloway, R., and P. Bebbington, 1987. The buffer theory of social support - a review of the literature. *Psychological Medicine* 17, 91-108.

Andreasen. N.C., 1984. *The Broken Brain, the Biological Revolution in Psychiatry.* Harper & Row, New York.

Baron, M., 1986a. Genetics of schizophrenia: I. Familial patterns and mode of inheritance. *Biological Psychiatry* 21, 1051-1066.

Baron, M., 1986b. Genetics of schizophrenia: II. Vulnerability traits and gene markers. *Biological Psychiatry* 21, 1189-1211.

Barondess, J.A., 1979. Disease and illness - a crucial distinction. *American Journal of Medicine* 66, 375-376.

Beatty, J., 1980. Optimal-design models and the strategy of model building in evolutionary biology. *Philosophy of Science* 47, 532-561.

Bechtel, W. (ed.), 1986. *Integrating Scientific Disciplines.* Nijhoff, Dordrecht.

Beckner, M., 1968. *The Biological Way of Thought.* University of California Press, Berkeley (second edition).

Begon, M., J.L. Harper and C.R. Townsend, 1986. *Ecology, Individuals, Populations and Communities.* Blackwell, Oxford.

Benor, D.J., 1984. Psychic healing. In: J.W. Salmon (ed.), *Alternative Medicines, Popular and Policy Perspectives,* pp. 165-190. Tavistock Publications, London.

Berg, J.H. van den, 1955. *The Phenomenological Approach to Psychiatry.* Thomas, Springfield.

Berg, J.H. van den, 1956. *Metabletica of Leer der Veranderingen.* Callenbach, Nijkerk (English translation 1962: *The Changeing Nature of Man.* Norton & Cy, New York).

Berg, J.H. van den, 1959. *Het Menselijk Lichaam, deel I.* Callenbach, Nijkerk.

Berg, J.H. van den, 1961. *Het Menselijk Lichaam, deel II*. Callenbach, Nijkerk.

Berg, J.H. van den, 1964. *De Psychiatrische Patiënt*. Callenbach, Nijkerk.

Berg, J.H. van den, 1965. *De Dingen*. Callenbach, Nijkerk.

Berg, J.H. van den, 1968. *Metabletica der Materie*. Callenbach, Nijkerk.

Berg, J.H. van den, 1978. A metabletic-philosophical evaluation of mental health. In: H.T. Engelhardt and S.F. Spicker (eds), *Mental Health: Philosophical Perspectives*, pp. 121-135. Reidel, Dordrecht.

Bertalanffy, L. von, 1949. *Das Biologische Weltbild*. Franck Verlag, Bern.

Bertalanffy, L. von, 1964. The mind-body problem: a new view. *Psychosomatic Medicine* 26, 29-45.

Bertalanffy, L. von, 1968. *General Systems Theory*. Braziller, New York.

Birdwhistell, R.L., 1970. *Kinesics and Context, Essays on Body Motion Communication*. University of Pennsylvania Press, Philadelphia.

Blaney, P.H., 1985. Stress and depression in adults: a critical review. In: T.M. Field, P.M. McCabe and N. Schneiderman (eds), *Stress and Coping*, pp. 263-283. Erlbaum, Hillsdale.

Bloch, G., 1985. *Body and Self, Elements of Human Biology, Behaviour, and Health*. Kauffman, Los Altos.

Bodman, F.H., 1968. Provers. *Journal of the American Institute of Homeopathy* 61, 79-88.

Bok, D., 1984. Needed: a new way to train doctors. *Harvard Magazine*, May, 32-43; June, 70-71.

Boorse, C., 1975. On the distinction between disease and illness. *Philosophy and Public Affairs* 5, 49-68.

Boorse, C. 1976. What a theory of mental health should be. *Journal of the Theory of Social Behaviour* 6, 61-84.

Boorse, C., 1977. Health as a theoretical concept. *Philosophy of Science* 44, 542-573.

Borst, C.V. (ed.), 1970. *The Mind/Brain Identity Theory*. Macmillan & Co., London.

Boss, M., 1971. *Grundriss der Medizin*. Huber, Bern.

Bradie, M., 1986. Assessing evolutionary epistemology. *Biology and Philosophy* 1, 401-459.

Braude, S.E., 1986. *The Limits of Influence. Psychokinesis and the Philosophy of Science*. Routledge and Kegan Paul, New York.

Bregman, L., 1982. *The Rediscovery of Inner Experience*. Nelson-Hall, Chicago.

Brody, H., 1985a. Placebo effect: an examination of Grünbaum's definition. In: L. White, B. Tursky and G.E. Schwartz (eds), *Pacebo. Theory,
Research, and Mechanisms*, pp. 37-58. Guilford Press, New York.

Brody, H., 1985b. Philosophy of medicine and other humanities: toward a
wholistic view. *Theoretical Medicine* 6, 243-255.

Brostoff, J., and S. Challacombe (eds), 1987. *Food Allergy and Intolerance*. Ballière Tindall, London.

Brown, W.M., 1985a. On defining 'disease'. *Journal of Medicine and
Philosophy* 10, 311-328.

Brown, W.M., 1985b. A critique of three conceptions of mental illness.
Journal of Mind and Behaviour 6, 553-576.

Buck, C., 1975. Popper's philosophy for epidemiologists. *International
Journal of Epidemiology* 4, 159-168.

Bunge, M., 1980. *The Mind-Body Problem, a Psychobiological Approach*.
Pergamon, Oxford.

Bunge, M., 1981. *Scientific Materialism*. Reidel, Dordrecht.

Buytendijk, F.J.J., 1948. *Algemene Theorie der Menselijke Houding en
Beweging*. Spectrum, Utrecht.

Buytendijk, F.J.J., 1965. *Prologomena van een Anthropologische Fysiolo-
gie*. Aula, Utrecht.

Canguilhem, G., 1966. *Le Normal et le Pathologique*. Presses Universi-
taires de France, Paris.

Carlson, R.J., 1975. *The End of Medicine*. Wiley, New York.

Carpenter, W.T., 1986. Thoughts on the treatment of schizophrenia.
Schizophrenia Bulletin 12, 527-539.

Cartwright, N., 1983. *How the Laws of Physics Lie*. Clarendon Press, Oxford.

Churchland, P.M., 1984. *Matter and Consciousness, a Contemporary Introduction to the Philosophy of Mind*. MIT Press, Cambridge Mass.

Churchland, P.M., 1985. Cognitive neurobiology: a computational hypothesis for laminar cortex. *Biology and Philosophy* 1, 25-61.

Churchland, P.M., and C. Hooker (eds), 1985. *Images of Science, Essays on
Realism and Empiricism*. University of Chicago Press, Chicago.

Churchland, Patricia S., 1986. *Neurophilosophy, toward a Unified Science
of the Mind-Brain*. MIT Press, Cambridge Mass.

Clark, A., 1980. *Psychological Models and Neural Mechanisms*. Clarendon
Press, Oxford.

Claxton, G., 1986. *Beyond Theory, the Impact of Eastern Religions on
Psychological Theory and Practice*. Wisdom Publications, London.

Cohen, D., and H. Cohen, 1986. Biological theories, drug treatments, and schizophrenia: a critical assessment. *Journal of Mind and Behaviour* 7, 11-36.

Craggs, M.D., and A.C. Carr, 1983. Neurophysiological aspects of psychiatry. In: M. Weller, ed., *The Scientific Basis of Psychiatry*, pp. 49-80. Baillière Tindall, London.

Culver, C., and B. Gert, 1982. *Philosophy in Medicine: Conceptual and Ethical Issues in Medicine and Psychiatry*. Oxford University Press, New York.

Cummins, R., 1983. *The Nature of Psychological Explanation*. MIT Press, Cambridge Mass.

Daniels, N., 1985. *Just Health Care*. Cambridge University Press, Cambridge.

Darden, L. and N. Maull, 1977. Interfield theories. *Philosophy of Science* 44, 43-64.

Davies, A.M., 1975. Comments on 'Popper's philosophy for epidemiologists' by Carol Buck, comment one. *International Journal of Epidemiology* 4, 169-170.

Dawkins, R., 1982. *The Extended Phenotype, the Gene as the Unit of Selection*. Freeman, Oxford.

Dawkins, R., and J.R. Krebs, 1979. Arms races between and within species. *Proceedings of the Royal Society, London*, B 205, 489-511.

Dennett, D.C., 1985. *Brainstorms*. Harvester Press, Brighton (original edition 1978).

Diekstra, R.F.W., and J.J. Dijkhuis, 1985. *Op de Psyche Gepast*. Ambo, Baarn.

Dijck, R. van, 1986. *Psychotherapie, Placebo en Suggestie*. Thesis, University of Leiden.

Dunning, A.J., 1980. Tusen Houston en Lourdes. In: A. Querido and J. Roos (eds), *Controversen in de Geneeskunde I*, pp. 216-222. Bunge, Utrecht.

Dyer, A.R., 1986. The concept of character: moral and therapeutic considerations. *British Journal of Medical Psychology* 59, 35-41.

Efron, D., 1941. *Gesture and Environment*. King's Crown Press, New York. (Reprinted in 1972 as *Gesture, Race and Culture*. Mouton, The Hague.)

Engel, B.T., 1985. Stress is a noun! No, a verb! No, an adjective! In: T.M. Field, P.M. McCabe, and N. Schneiderman (eds), *Stress and Coping*, pp. 3-12. Erlbaum, Hillsdale.

Engel, B.T., 1986. Psychosomatic medicine, behavioral medicine, just plain medicine. *Psychosomatic Medicine* 48, 466-479.

Engel, G.L., 1977. The need for a new medical model: a challenge for biomedicine. *Science* 196, 129-136.

Engel, G.L., 1981. The clinical application of the biopsychosocial model. *Journal of Medicine and Philosophy* 6, 101-123.

Engelhardt, H.T., 1973. *Mind-Body, a Categorial Relation.* Nijhoff, The Hague.

Engelhardt, H.T., 1981. The concepts of health and disease. In: A.L. Caplan, H.T. Engelhardt and J.J. McCartney (eds), *Concepts of Health and Disease,* pp. 31-46. Addison-Wesley, London (reprint, original article 1975).

Engelhardt, H.T., 1984. Clinical problems and the concept of disease. In: L. Nordenfelt and B.I.B. Lindahl (eds), *Health, Disease, and Causal Explanations in Medicine,* pp. 27-41. Reidel, Dordrecht.

Evans, F.J., 1985. Expectancy, therapeutic instructions, and the placebo response. In: L. White, B. Tursky and G.E. Schwartz (eds), *Placebo. Theory, Research, and Mechanisms,* pp. 215-228. Guilford Press, New York.

Eysenck, H.J. and C. Sargent, 1982. *Explaining the Unexplained, Mysteries of the Paranormal.* Weidenfeld and Nicholson, London.

Eysenck. H.J., 1983. Stress, disease, and personality: the 'inoculation effect'. In: C.L. Cooper (ed.), *Stress Research,* pp. 121-146. Wiley, Chichester.

Farber, M., 1967. *Phenomenology and Existence.* Harper, New York.

Faust, D., 1984. *The Limits of Scientific Reasoning.* University of Minnesota Press, Minneapolis.

Feinstein, A.R., 1967. *Clinical Judgement.* Williams and Wilkins, Baltimore.

Feinstein, A.R., 1985. *Clinical Epidemiology: The Architecture of Clinical Research.* Saunders, Philadelphia.

Feinstein, A.R., 1987. The intellectual crisis in clinical sciences: medaled models and muddled mettle. *Perspectives in Biology and Medicine* 30, 215-230.

Feyerabend, P.K., 1975. *Against Method.* NLB, London.

Feyerabend, P.K., 1978. *Science in a Free Society.* NLB, London.

Field, T.M., P.M. McCabe and N. Schneiderman (eds), 1985. *Stress and Coping.* Erlbaum, Hillsdale.

Flanagan, O.J., 1984. *The Science of the Mind*. MIT Press, Cambridge Mass.

Fraassen, B.C. van, 1980. *The Scientific Image*. Clarendon Press, London.

Freidson, E., 1976. *Profession of Medicine. A Study of the Sociology of Applied Knowledge*. Harper & Row, New York.

Gabbard, G.O., and S.W. Twemlow, 1984. *With the Eyes of the Mind, an Empirical Analysis of Out-of-Body States*. Praeger, New York.

Garfield, S.L., 1984. Psychotherapy: efficacy, generality and specificity. In: J.B.W. Williams and R.L. Spitzer (eds), *Psychotherapy Research: Where are We and Where Should We Go*, pp. 295-305. Guilford Press, New York.

Ghiselin, M.T., 1987. Species concepts, individuality, and objectivity. *Biology and Philosophy* 2, 127-143.

Gibson, R.G., S.L.M. Gibson, A.D. McNeill, and W.W. Buchanan, 1980. Homeopathic therapy in rheumatoid arthritis: evaluation by double-blind clinical therapeutic trial. *British Journal of Clinical Pharmacology* 9, 453-459.

Goldman, H.H. (ed.), 1984. *Review of General Psychiatry*. Lange Medical Publications, Los Altos.

Goldman, H.H., and S.A. Forman, 1984. Psychiatric diagnosis and psychosocial formulation. In: H.H. Goldman (ed.), *Review of General Psychiatry*, pp. 139-150. Lange Medical Publications, Los Altos.

Goldstein, M.J., 1987. Psychosocial issues. *Schizophrenia Bulletin* 13, 157-171.

Goosens, W.K., 1980. Values, health, and medicine. *Philosophy of Science* 47, 100-115.

Gorovitz, S., and A. MacIntyre, 1976. Toward a theory of medical fallibility. *Journal of Medicine and Philosophy* 1, 51-71.

Graham, D.T., 1979. What place in medicine for psychosomatic medicine? *Psychosomatic Medicine* 41, 357-367.

Gräsbeck, R., 1984. Health and disease from the point of view of the clinical laboratory. In: L. Nordenfelt and B.I.B. Lindahl (eds), *Health, Disease, and Causal Explanations in Medicine*, pp. 47-60. Reidel, Dordrecht.

Greyson, B. and C.P. Flynn (eds), 1984. *The Near-Death Experience, Problems, Prospects, Perspectives*. Thomas, Springfield.

Griggs, W.B., 1968. A proving of indol. *Journal of The American Institute of Homeopathy* 61, 89-95.

Groen, J., 1957. Psychosomatic disturbances as a form of substituted behaviour. *Journal of Psychosomatic Research* 2, 85-96.

Grünbaum, A., 1984. *The Foundations of Psychoanalysis: a Philosophical Critique*. University of California Press, Berkeley.

Grünbaum, A., 1985. Explication and implications of the pacebo concept. In: L. White, B. Tursky and G.E. Schwartz (eds), *Placebo. Theory, Research and Mechanisms,* pp. 3-36. Guilford Press, New York.

Grünbaum, A., 1986a. The placebo concept in medicine and psychiatry. *Psychological Medicine* 16, 19-38.

Grünbaum, A., 1986b. Précis of the foundations of psychoanalysis: a critique (followed by peer reviews). *Behavioral and Brain Sciences* 9, 217-284.

Habermas, J., 1971. *Knowledge and Human Interests.* (Translated from German.) Beacon Press, Boston.

Hacking, I., 1983. *Representing and Intervening.* Cambridge University Press, Cambridge.

Halbreich, U. (ed.), 1987. *Hormones and Depression.* Raven Press, New York.

Hamilton, V., 1982. Cognition and stress: an information processing model. In: L. Goldberger and S. Breznitz, eds, *Handbook of Stress, Theoretical and Clinical Aspects,* pp. 105-120. Free Press, New York.

Hanly, C., 1985. Logical and conceptual problems of existential psychiatry. *Journal of Nervous and Mental Disease* 173, 263-275.

Harding, S.G., 1976. *Can Theories be Refuted?* Reidel, Dordrecht.

Have, H. ten, J. Bergsma, and J. Broekman, 1987. Preface [to special issue]. *Theoretical Medicine* 8, 100-103.

Heelan, P.A., 1987. Husserl's later philosophy of natural science. *Philosophy of Science* 54, 368-390.

Heil, J., 1986. Formalism and psychological explanation. *Journal of Mind and Behavior* 7, 1-10.

Hempel, C.G., 1965. *Aspects of Scientific Explanation and Other Essays in the Philosophy of Science.* Free Press, New York.

Hempel, C.G., and P. Oppenheim, 1948. Studies in the logic of explanation. *Philosophy of Science* 15, 135-175.

Hertogh, C., 1987. Life and the scientific concept of life. *Theoretical Medicine* 8, 117-126.

Higgins, P.M., 1984. Things aren't what they seem. *Journal of the Royal Society of Medicine* 77, 728-737.

Hinkles, L.E., 1977. The concept of "stress" in the biological and the social sciences. In: Z.J. Lipowski, D.R. Lipsitt and P.C. Whybrow (eds),

Psychosomatic Medicine, Current Trends and Applications, pp. 27-49. Oxford University Press, New York.

Hollis, M., and S. Lukes (eds), 1982. *Rationality and Relativism.* Blackwell, Oxford.

Hook, S. (ed.), 1960. *Dimensions of Mind.* Collier-Macmillan, London.

Hull, D.L., 1965. The effect of essentialism on taxonomy: two thousand years of stasis, parts I and II. *British Journal for the Philosophy of Science* 15, 314-326 and 16, 1-18.

Hull, D.L., 1968. The operational imperative: sense and nonsense in operationism. *Systematic Zoology* 17, 438-457.

Hull, D.L., 1981. Reduction in genetics. *Journal of Medicine and Philosophy* 6, 125-143.

Hunter, J.O. and V.A. Jones, 1985. *Food and the Gut.* Baillière Tindall, London.

Husserl, E., 1976. *Die Krisis der Europäischen Wissenschaften und die Transzendentale Phänomenologie.* Nijhoff, The Hague. (Second edition edited by W. Biemel.)

Illich, I., 1975. *Medical Nemesis - the Expropriation of Health.* Boyars, London.

Jablensky, A., 1987. Multicultural studies and the nature of schizophrenia: a review. *Journal of the Royal Society of Medicine* 80, 162-167.

Jacobs, M., 1968. The shifting existence of Western man: an introduction to J.H. van den Berg's work on metabletics, with a summary of his investigations of the metabletics of the human body. *Humanitas, Journal of the Institute of Man,* IV, 25-73.

Jacobs, M., 1969. Metabletics of loneliness: an account of J.H. van den Berg's *Life in Multiplicity. Social Research* 36, 608-639.

Jacobsen, M., 1976. Against Popperian epidemiology. *International Journal of Epidemiology* 5, 9-11.

Jennings, D., 1986. The confusion between disease and illness in clinical medicine. *Canadian Medical Association Journal* 135, 865-870.

Jores, A., 1966. *Die Medizin in der Krise Unserer Zeit.* Huber, Bern.

Karasu, T.B., 1986. The specificity versus nonspecificity dilemma: Toward identifying therapeutic change agents. *American Journal of Psychiatry* 143, 687-695.

Kasl, S.V., 1983. Pursuing the link between stressful life experiences and disease: a time for reappraisal. In: C.L. Cooper (ed.), *Stress Research,* pp. 79-102. Wiley, Chichester.

Kass, L.R., 1975. Regarding the end of medicine and the pursuit of health. *Public Interest* 40, 11-42.

Kendell, R.E., 1975. *The Role of Diagnosis in Psychiatry*. Blackwell, Oxford.

Kendell, R.E., 1983. The principles of classification in relation to mental disease. In: M Shepherd and O.L. Zangwill (eds), *Handbook of Psychiatry I, General Psychopathology,* pp. 191-198. Cambridge University Press, Cambridge.

Kendell, R.E., 1984. Reflections on psychiatric classification - For the architects of DSM-III and ICD-10. *Integrative Psychiatry* 2, 43-57.

Kienle, G., and R. Burkhardt, 1983. *Der Wirksamkeits-Nachweis für Arzneimittel, Analyse einer Illusion*. Urachhaus, Stuttgart.

Kim, J., 1978. Supervenience and nomological incommensurability. *American Philosophical Quarterly* 15, 149-156.

Kincaid, H., 1986. Reduction, explanation and individualism. *Philosophy of Science* 53, 492-513.

Kitcher, P., 1984. Species. *Philosophy of Science* 51, 308-333.

Klein, D.F., and J.G. Rabkin, 1984. Specificity and strategy in psychotherapy research and practice. In: J.B.W. Williams and R.L. Spitzer (eds), *Psychotherapy Research: Where are We and Where Should We Go,*pp. 306-331. Guilford Press, New York.

Kleinman, A., 1980. *Patients and Healers in the Context of Culture*. University of California Press, Berkeley.

Kleinman, A., 1984. Indigeneous systems of healing: questions for professional, popular, and folk care. In: J.W. Salmon (ed.), *Alternative Medicines, Popular and Policy Perspectives,* pp. 138-164. Tavistock Publications, London.

Kleinman, A., and B. Good, 1985. *Culture and Depression. Studies in the Anthropology and Cross-Cultural Psychiatry of Affect and Disorder*. University of California Press, Berkeley.

Kopelman, L., and J. Moskop, 1981. The holistic health movement: a survey and critique. *Journal of Medicine and Philosophy* 6, 209-235.

Kosa, J., and I.K. Zola (eds), 1975. *Poverty and Health, a Sociological Analysis*. Harvard University Press, Cambridge Mass.

Krebs, C.J., 1972. *Ecology, the Experimental Analysis of Distribution and Abundance*. Harper & Row, New York.

Kuhn, T., 1962. *The Structure of Scientific Revolutions*. University of Chicago Press, Chicago.

Lacey, J.I., D.E. Bateman, and R. van Lehn, 1953. Autonomic response specificity, an experimental study. *Psychosomatic Medicine* 15, 8-21.

Lakatos, I., 1972. Falsification and the methodology of research programs. In: I. Lakatos and A. Musgrave (eds), *Criticism and the Growth of Knowledge* (revised edition), pp. 91-196. Cambridge University Press, Cambridge.

LaPatra, J., 1978. *Healing*. McGraw-Hill, New York.

Lazare, A., 1981. Hidden conceptual models in clinical psychiatry. In: A.L. Caplan, H.T. Engelhardt and J.J. McCartney (eds), *Concepts of Health and Disease,* pp. 419-431. Addison-Wesley, London (reprint; original article 1972).

Lazarus, R.S., 1966. *Psychological Stress and the Coping Process.* McGraw-Hill, New York.

Lazarus, R.S., 1977. Psychological stress and coping in adaptation and illness. In: Z.J. Lipowski, D.R. Lipsitt and P.C. Whybrow (eds), *Psychosomatic Medicine, Current Trends and Applications,* pp. 14-26. Oxford University Press, New York.

Leder, D., 1984. Medicine and paradigms of embodiment. *Journal of Medicine and Philosophy* 9, 29-43.

Ledoux, J.E., and W. Hirst, 1986. *Mind and Brain. Dialogues in Cognitive Neuroscience.* Cambridge University Press, Cambridge.

Leplin, J., 1986. Methodological realism and scientific rationality. *Philosophy of Science* 53, 31-51.

LeShan, L., 1980. *Clairvoyant Reality, towards a General Theory of the Paranormal.* Turnstone Press, Wellingborough.

Levenson, R.W., 1979. Effects of thematically relevant and general stressors on specificity of responding in asthmatic and nonasthmatic subjects. *Psychosomatic Medicine* 41, 28-38.

Lewith, G.T., and J.N. Kenyon, 1985. *Clinical Ecology, The Treatment of Ill-Health Caused by Environmental Factors.* Thorsons Publishers, Wellingborough.

Lewontin, R.C., 1974. *The Genetic Basis of Evolutionary Change.* Columbia University Press, New York.

Lipowski, Z.J., 1984. What does the word "psychosomatic" really mean? A historical and semantic inquiry. *Psychosomatic Medicine* 46, 153-169.

Lusted, L.B., 1968. *Introduction to Clinical Decision Making.* Thomas, Springfield.

Lüth, P., 1976. *Medizin als Politik.* Luchterhand, Darmstadt.

MacIntyre, A., 1981. *After Virtue*. Notre Dame University Press, South Bend.

Macklin, R., 1981. Mental health and mental illness: some problems of definition and concept formation. In: A.L. Caplan, H. Engelhardt and J.J. McCartney (eds), *Concepts of Health and Disease*, pp. 392-418. Addison-Wesley, London (reprint, original article 1972).

Maclure, M., 1985. Popperian refutation in epidemiology. *American Journal of Epidemiology* 121, 343-350.

Maclure, M., 1986. Criticism and the growth of epidemiologic knowledge (re: "Popperian refutation in epidemiology"). *American Journal of Epidemiology* 123, 1119-1121.

Maclure, M., 1987. On the logic and practice of epidemiology. *American Journal of Epidemiology* 126, 554.

Maddocks, I., 1985. Alternative Medicine. *Medical Journal of Australia* 142, 547-551.

Malan, D.H., 1976. *Toward the Validation of Dynamic Psychotherapy*. Plenum Press, London.

Malcolm, N., 1968. The conceivability of mechanism. *Philosophical Review* 77, 45-72.

Malmo, R.B., and C. Shagass, 1949. Physiological study of symptom mechanisms in psychiatric patients under stress. *Psychosomatic Medicine* 11, 25-29.

Mandler, G., 1982. Stress and thought processes. In: L. Goldberger and S. Breznitz (eds), *Handbook of Stress, Theoretical and Clinical Aspects*, pp. 88-104. Free Press, New York.

Maretzki, T.W., and E. Seidler, 1985. Biomedicine and naturopathic medicine in West Germany, a historical ethnomedical view of a stormy relationship. *Culture, Medicine and Psychiatry* 9, 383-422.

Margulies. A.M., 1984. Toward empathy: the uses of wonder. *American Journal of Psychiatry* 141, 1025-1033.

Marmot, M., 1976. Facts, opinions and affaires du coeur. *American Journal of Epidemiology* 6, 519-526.

Marmot, M., 1986. Epidemiology and the art of the soluble. *Lancet* April 19, 897-900.

Maslow, A.H., 1954. *Motivation and Personality*. Harper and Row, New York.

Maslow, A.H., 1962. *Toward a Psychology of Being*. Van Nostrand, New York.

Mayr, E., 1969. *Principles of Systematic Zoology*. McGraw-Hill, New York.

Mayr, E., 1982. *The Growth of Biological Thought*. Harvard University Press, Cambridge Mass.

Mayr, E., 1987. The ontological status of species: scientific progress and philosophical terminology. *Biology and Philosophy* 2, 145-166.

McGuire, M.T., 1986. Phenomenological classification systems: the case of DSM-III. *Perspectives in Biology and Medicine* 30, 135-147.

McKeown, T., 1983. A basis for health strategies. *British Medical Journal* 287, 594-596.

McKeown, T., 1984. *The Role of Medicine*. Blackwell, Oxford (2nd ed.).

McWhinney, I.R., 1986. Are we on the brink of a major transformation of clinical method? *Canadian Medical Association Journal* 135, 873-878.

Mechanic, D., 1986. The concept of illness behaviour: culture, situation and predisposition. *Psychological Medicine* 16, 1-7.

Merleay-Ponty, M., 1945. *Phénoménologie de la Perception*. Gaillimard, Paris.

Merleau-Ponty, M., 1961. L'oeil et l'esprit. *Les Temps Modernes* 17, 193-227.

Merleau-Ponty, M., 1964. *Le Visible et l'Invisible*. Gaillimard, Paris.

Michaelson. M.G., 1981. The coming medical war. *New York Review of Books,* July 1.

Mikhail, A., 1985. Stress, a psychophysiological conception. In: A. Monat and R.S. Lazarus (eds). *Stress and Coping, an Anthology,* pp. 30-39. Columbia University Press, New York (reprint; original article 1981).

Miller, J.G., 1978. *Living Systems*. McGraw-Hill, New York.

Miller, R., 1981. *Meaning and Purpose in the Intact Brain, a Philosophical, Psychological, and Biological Account of Conscious Processes*. Clarendon Press, Oxford.

Mirsky, A.F., and C.C. Duncan, 1986. Etiology and expression of schizophrenia: neurobiological and psychosocial factors. *Annual Review of Psychology* 37, 291-319.

Molen, M.W. van der, and J.F. Orlebeke, 1980. Phasic heart rate and the U-shaped relationship between choice reaction time and auditory signal intensity. *Psychophysiology* 17, 471-481.

Monat, A., and R.S. Lazarus (eds), 1985. *Stress and Coping, an Anthology*. Columbia University Press, New York.

Moody, R.A., 1975. *Life after Life*. Mockingbird, Covington.

Moore, M.S., 1984. *Law and Psychiatry, Rethinking the Relationship.* Cambridge University Press, Cambridge.

Mössinger, P., 1984. *Homöopathie und Naturwissenschaftliche Medizin, zur Ueberwindung der Gegensätze.* Hippokrates, Stuttgart.

Munson, R., 1981. Why medicine cannot be a science. *Journal of Medicine and Philosophy* 6, 183-208.

Muscari, P.G., 1981. The structure of mental disorder. *Philosophy of Science* 48, 553-572.

Nagel, E., 1961. *The Structure of Science.* Routledge, London.

Nagel, T., 1986. *The View from Nowhere.* Oxford University Press, New York.

Natsoulas, T., 1986. On the radical behaviorist conception of consciousness. *Journal of Mind and Behavior* 7, 87-115.

Navarro, V., 1980. Work, ideology and sciences: the case of medicine. *Social Science and Medicine* 14C, 191-205.

Neu, J., 1977. *Emotion, Thought and Therapy.* Routledge & Kegan Paul, London.

Ney, P.G., C. Collins, and C. Spensor, 1986. Double blind: double talk or are there ways to do better research. *Medical Hypotheses* 21, 119-126.

Newton-Smith, W.H., 1981. *The Rationality of Science.* Routledge & Kegan Paul, Boston.

Nordenfelt, L., and B.I.B. Lindahl (eds), 1984. *Health, Disease, and Causal Explanation in Medicine.* Reidel, Dordrecht.

O'Connor, J. (ed.), 1969. *Modern Materialism: Readings on Mind-Body Identity.* Harcourt, Brace & World, New York.

Orlebeke, J.F., M. van der Molen, R.J.M. Somsen, and L.J.P. van Doornen, 1985. Individual response specificity in phasic cardiac activity, implications for stress research. In: C. Spielberger and I. Sarason (eds), *Stress and Anxiety,* pp. 163-175. McGraw Hill, New York.

Oyama, S., 1985. *The Ontogeny of Information, Developmental Systems and Evolution.* Cambridge University Press, Cambridge.

Ozar, D.T., 1983. What should count as basic health care? *Theoretical Medicine* 4, 129-141.

Paré, W.P., and G.B. Glavin, 1986. Restraint stress in biomedical research: a review. *Neuroscience and Biobehavioral Reviews* 10, 339-370.

Parsons, T., 1981. Definitions of health in the light of American values and social structure. In: A.L. Caplan, H.T. Engelhardt and J.J. McCartney (eds), *Concepts of Health and Disease,* pp. 57-81. Addison-Wesley, London (reprint, original article 1958).

Paul, G.L., 1985. Can pregnancy be a placebo effect? Terminology, designs, and conclusions in the study of psychosocial and pharmacological treatments of behavioral disorders. In: L. White, B. Tursky and G.E. Schwartz (eds), *Placebo. Theory, Research and Mechanisms,* pp. 137-166. Guilford Press, New York.

Pellegrino, E., and D. Thomasma, 1981. *A Philosophical Basis of Medical Practice.* Oxford University Press, Oxford.

Pelletier, K.R., 1979. *Holistic Medicine.* Delacorte Press, New York.

Peterfreund, E., 1983. *The Process of Psychoanalytic Therapy, Models and Strategies.* The Analytic Press, distributed by Erlbaum, Hillsdale.

Polanyi, M., 1958. *Personal Knowledge, towards a Post-Critical Philosophy.* University of Chicago Press, Chicago.

Polanyi, M., and H. Prosch, 1975. *Meaning.* University of Chicago Press, Chicago.

Popper, K.R., 1934. *Logik der Forschung.* Springer, Wien (English translation 1959: *The Logic of Scientific Discovery,* Hutchinson, London).

Popper, K.R., 1963. *Conjectures and Refutations, the Growth of Scientific Knowledge.* Harper and Row, New York.

Popper, K.R., and J.C. Eccles, 1981. *The Self and Its Brain.* Springer, Berlin (original edition 1977).

Putnam, H., 1981. *Reason, Truth and History.* Cambridge University Press, Cambridge.

Quine, W.V.O., 1953. *From a Logical Point of View.* MIT Press, Cambridge Mass.

Quine, W.V.O., 1960. *Word and Object.* MIT Press, Cambridge Mass.

Rachlin, H., 1985. Pain and behavior. *Behavioral and Brain Sciences* 8, 43-83.

Rack, P. (ed.), 1982. *Race, Culture, and Mental Disorder.* Tavistock Publications, London.

Ragsdale, J.D., and C. Fry Silvia, 1982. Distribution of kinesic hesitation phenomena in spontaneous speech. *Language and Speech* 25, 185-190.

Rahimtoola, S.H., 1982. Coronary bypass surgery for chronic angina, a perspective. *Circulation* 65, 225-241.

Reilly, D.T., C. McSharry, M.A. Taylor and T. Atchinson, 1986. Is homoeopathy a placebo response? Controlled trial of homoeopathic potency, with pollen in hayfever as a model. *Lancet* 18 October, 881-886.

Reiser, M.F., 1984. *Mind, Brain, Body, toward a Convergence of Psychoanalysis and Neurobiology.* Basic Books, New York.

Rescher, N., 1987. *Scientific Realism, a Critical Reappraisal.* Reidel, Dordrecht.

Reus, V.I., 1984. Affective disorders. In: H.H. Goldman (ed.), *Review of General Psychiatry,* pp. 346-361. Lange Medical Publications, Los Altos.

Richards, R.J., 1986. A defense of evolutionary ethics. *Biology and Philosophy* 1, 265-293.

Richters, A., and E. Bonsel, 1987. The judgement comes with healing in its wings: a call for rational detachment. *Theoretical Medicine* 8, 147-162.

Ricoeur, P., 1981. *Hermeneutics and the Human Sciences.* (Translated from French.) Cambridge University Press, Cambridge.

Roe, W., 1984. "Science" in the practice of medicine: its limitations and dangers, as exemplified by a study of the natural history of acute bronchial asthma in children. *Perspectives in Biology and Medicine* 27, 386-400.

Rose, M.R., 1985. The evolution of senescence. In: P.J. Greenwood, P.H. Harvey and M. Slatkin (eds), *Essays in Honour of John Maynard Smith,* pp. 117-128. Cambridge University Press, Cambridge.

Rosenberg, A., 1978. The supervenience of biological concepts. *Philosophy of Science* 45, 368-386.

Rosenberg, A., 1980. *Sociobiology and the Preemption of Social Science.* Johns Hopkins University Press, Baltimore.

Rosenberg, A., 1985. *The Structure of Biological Science.* Cambridge University Press, Cambridge.

Rosenthal, D.M. (ed.), 1971. *Materialism and the Mind-Body Problem.* Prentice-Hall, Englewood Cliffs.

Rosenthal, R., 1985. Designing, analyzing, interpreting, and summarizing placebo studies. In: L. White, B. Tursky and G.E. Schwartz (eds), *Placebo. Theory, Research, and Mechanisms,* pp.110-136. Guilford Press, New York.

Rouse, J., 1987. Husserlian phenomenology and scientific realism. *Philosophy of Science* 54, 222-232.

Ruse, M., 1981. Are homosexuals sick? In: A.L. Caplan, H.T. Engelhardt and J.J. McCartney (eds), *Concepts of Health and Disease,* pp. 693-723. Addison-Wesley, London.

Ruse, M., 1986. *Taking Darwin Seriously.* Blackwell, Oxford.

Russell, J., 1984. *Explaining Mental Life, Some Philosophical Issues in Psychology.* Macmillan, London.

Rijnberk, G. van, 1942. *Inleiding tot de Studie der Geneeskunde.* Servire, Den Haag.

Sabom, M.B., 1982. *Recollections of Death: a Medical Investigation.* Harper & Row, New York.

Sacks, A.D., 1983. Nuclear magnetic resonance spectra of homeopathic remedies. *Journal of Holistic Medicine* 5, 172-177.

Salmon, W.C., 1984. *Scientific Explanation and the Causal Structure of the World.* Princeton University Press, Princeton.

Sassower, R., and M.A. Grodin, 1987. Scientific uncertainty and medical responsibility. *Theoretical Medicine* 8, 221-234.

Schaffner, K.F., 1967. Approaches to reduction. *Philosophy of Science* 34, 137-147.

Schaffner, K.F., 1976. Reduction in biology: prospects and problems. In: R.S. Cohen et al. (eds), *Boston Studies in the Philosophy of Science* XXXII, pp. 613-632. Reidel, Dordrecht.

Schaffner, K.F., 1986. Exemplar reasoning about biological models and diseases: a relation between the philosophy of medicine and philosophy of science. *Journal of Medicine and Philosophy* 11, 63-80.

Scheflen, A., and A.E. Scheflen, 1972. *Body Language and the Social Order.* Prentice Hall, Englewood Cliffs.

Scheflen, A.E., 1973. *Communicational Structure: Analysis of a Psychotherapy Transaction.* Indiana University Press, Bloomington.

Schroeder, S.A., 1984a. A comparison of Western European and U.S. university hospitals. *Journal of the American Medical Association* 252, 240-246.

Schroeder, S.A., 1984b. Western European responses to physician oversupply. Lessons for the United States. *Journal of the American Medical Association* 252, 373-384.

Schwartz, M.A., and O.P. Wiggins, 1985. Science, humanism, and the nature of medical practice: a phenomenological view. *Perspectives in Biology and Medicine* 28, 331-361.

Schwartz, M.A., and O.P. Wiggins, 1986a. Logical empiricism and psychiatric classification. *Comprehensive Psychiatry* 27, 101-114.

Schwartz, M.A., and O.P. Wiggins, 1986b. Systems and the structuring of meaning: contributions to a biopsychosocial medicine. *American Journal of Psychiatry* 143, 1213-1221.

Schwartz, S.R., and B. Africa, 1984. Schizophrenia disorders. In: H.H. Goldman (ed.), *Reviews of General Psychiatry,* pp. 311-327. Lange Medical Publications, Los Altos.

Scofield, A.M., 1984. Experimental research in homeopathy - a critical review. *British Homeopathic Journal* 73, 161-226.

Selye, H., 1936. A syndrome produced by diverse nocuous agents. *Nature* 138, 32.

Selye, H., 1950. *The Psychology and Pathology of Exposure to Stress.* Acta, Montreal.

Selye, H., 1983. The stress concept: past, present and future. In: C.L. Cooper (ed.), *Stress Research, Issues for the Eighties,* pp. 1-20. Wiley, Chichester.

Siegel, H., 1985. What is the question concerning the rationality of science? *Philosophy of Science* 52, 517-537.

Silver, G., 1986. Whom do we serve? *Lancet* February 8, 315-316.

Sloep, P.B., and W.J. van der Steen, 1987. The nature of evolutionary theory: the semantic challenge. *Biology and Philosophy* 2, 1-15.

Smith, A., 1975. Comments on 'Popper's philosophy for epidemiologists' by Carol Buck, comment two. *International Journal of Epidemiology* 4, 171-172.

Smith, P., and O.R. Jones, 1986. *The Philosophy of Mind, an Intro-duction.*Cambridge University Press, Cambridge.

Sneath, P.H.A., and R.R. Sokal, 1973. *Numerical Taxonomy.* Freeman, San Francisco.

Sneed, J.D., 1971. *The Logical Structure of Mathematical Physics.* Reidel, Dordrecht.

Sperry, R.W., 1969. A modified concept of consciousness. *Psychological Review* 76, 532-536.

Sperry, R.W., 1976. Mental phenomena as causal determinants in brain function. In: G.G. Globus, G. Maxwell and I. Savodnik (eds), *Consciousness and the Brain,* pp. 163-177. Plenum, New York.

Sperry, R.W., 1983. *Science and Moral Priority.* Basil Blackwell, Oxford.

Spiegelberg, H., 1960. *The Phenomenological Movement.* Nijhoff, The Hague.

Staal, F., 1975. *Exploring Mysticism.* Penguin, New York.

Stalker, D., and C.G. Glymour (eds), 1985. *Examining Holistic Medicine.* Prometheus Books, Buffalo.

Steen, W.J. van der, 1982. *Algemene Methodologie voor Biologen.* Bohn, Scheltema and Holkema, Utrecht.

Steen, W.J. van der, 1986a. Methodological problems in evolutionary biology. V. The impact of supervenience. *Acta Biotheoretica* 35, 185-191.

Steen, W.J. van der, 1986b. Methodological problems in evolutionary biology. VI. The force of evolutionary epistemology. *Acta Biotheoretica* 35, 193-204.

Steen, W.J. van der, and B. Voorzanger, 1984. Sociobiology in perspective. *Journal of Human Evolution* 13, 25-32.

Steen, W.J. van der, and B. Voorzanger, 1986. Methodological problems in evolutionary biology. VII. The species plague. *Acta Biotheoretica* 35, 205-221.

Stegmüller, W., 1969. *Probleme und Resultate der Wissenschaftstheorie und Analytischen Philosophie. I. Wissenschaftliche Erklärung und Begründung.* Springer, Berlin.

Stegmüller, W., 1973. *Probleme und Resultate der Wissenschaftstheorie und Analytischen Philosophie. II. Theorie und Erfahrung.* Springer, Berlin.

Stein, M., 1986. A reconsideration of specificity in psychosomatic medicine: from olfaction to the lymphocyte. *Psychosomatic Medicine* 48, 3-22.

Stenseth, N.C., 1985. Darwinian evolution in ecosystems: the red queen view. In: P.J. Greenwood, P.H. Harvey and M. Slatkin (eds), *Evolution, Essays in Honour of John Maynard Smith,* pp. 55-72. Cambridge University Press, Cambridge.

Stephens, G.G., 1986. Careers of illness: problems in the diagnosis of chronic illness. *Perspectives in Biology and Medicine* 29, 464-474.

Suppe, F., 1977. *The Structure of Scientific Theories.* University of Illinois Press, Chicago (2nd ed.).

Susser, M., 1986. The logic of sir Karl Popper and the practice of epidemiology. *American Journal of Epidemiology* 124, 711-718.

Susser, M., 1987. Reply. *American Journal of Epidemiology* 126, 554-555.

Thomasma, D.C., and E.D. Pellegrino, 1987. Challenges for a philosophy of medicine of the future: a response to fellow philosophers in the Netherlands. *Theoretical Medicine* 8, 187-204.

Thompson, J.B., 1981. *Critical Hermeneutics, a Study in the Thought of Paul Ricoeur and Jürgen Habermas.* Cambridge University Press, Cambridge.

Thompson, P., 1983. The structure of evolutionary theory: a semantic approach. *Studies in the History and Philosophy of Science* 14, 215-229.

Thung, P.J., 1957. The relation between amyloid and ageing in comparative gerontology. *Gerontologia* 1, 260-279.

Thung, P.J., 1962. Ageing in the gonad-adrenal system. *Gerontologia* 6, 41-64.

Thung, P.J., 1964. Ontwikkeling en ouderdom, biologisch gezien. In: J. Huyts et al., *Ouderdom en Ontwikkeling,* pp. 73-96. Brand, Hilversum.

Thung, P.J., 1965. Over de zogenaamde ouderdom. *Huisarts en Wetenschap* 8, 19-27.

Thung, P.J., 1980. Alternatieve geneeswijzen: achtergronden en definities. In: A. Querido and J. Roos (eds), *Controversen in de Geneeskunde I,* pp. 84-97. Bunge, Utrecht.

Tiemersma, D., 1987. Ontology and ethics in the foundation of medicine and the relevance of Levinas' view. *Theoretical Medicine* 8, 127-133.

Trowell, H.C., and D.P. Burkitt, 1981. *Western Diseases: Their Emergence and Prevention.* Arnold, London.

Ullrich, A.C., 1984. Traditional healing in the third world. *Journal of Holistic Medicine* 6, 200-212.

Underwood, G., 1982. *Aspects of Consciousness, Vol. 3, Awareness and Self-Awareness.* Academic Press, London.

Underwood, G., and R. Stevens, 1979. *Aspects of Consciousness, Vol. 1, Psychological Issues.* Academic Press, London.

Underwood, G., and R. Stevens, 1981. *Aspects of Consciousness, Vol. 2, Structural Issues.* Academic Press, London.

Valen, L. van, 1973. A new evolutionary law. *Evolutionary Theory* 1, 1-30.

Verwey, G., 1987. Toward a systematic philosophy of medicine. *Theoretical Medicine* 8, 163-177.

Vithoulkas, G., 1980. *The Science of Homeopathy.* Grove Press, New York.

Voorzanger, B., 1987a. Methodological problems in evolutionary biology. VIII. Biology and culture. *Acta Biotheoretica* 36, 23-34.

Voorzanger, B., 1987b. *Woorden, Waarden, en de Evolutie van Gedrag. Humane Sociobiologie in Methodologisch Perspectief.* Free University Press, Amsterdam.

Waddell, G., M. Bircher, D. Finlayson, and C.J. Main, 1984. Symptoms and signs: physical disease or illness behaviour? *British Medical Journal* 289, 739-741.

Waichbroit, R., 1987. Theories of rationality and principles of charity. *British Journal for the Philosophy of Science* 38, 35-47.

Wallach, M.A., and L. Wallach, 1983. *Psychology's Sanction for Selfishness, the Error of Egoism in Theory and Therapy.* Freeman, San Francisco.

Wartofsky, M.W., 1986. Clinical judgement, expert programs, and cognitive style: a counter-essay in the logic of diagnosis. *Journal of Medicine and Philosophy* 11, 81-92.

Watts, F.N. (ed.), 1985. *New Developments in Clinical Psychology.* British Psychological Society in association with Wiley, Chichester.

Waxler, N.E., 1984. Behavioral convergence and institutional separation: an analysis of plural medicine in Sri Lanka. *Culture, Medicine and Psychiatry* 8, 187-205.

Weed, D.L., and B.J. Trock, 1986. Criticism and the growth of epidemiologic knowledge. (Re: "Popperian refutation in epidemiology"). *American Journal of Epidemiology* 123, 1119-1120.

Weiner, H., 1982. The prospects for psychosomatic medicine: selected topics. *Psychosomatic Medicine* 44, 491-512.

Weinstein, M.C., and V. Fineberg, 1980. *Clinical Decision Analysis.* Saunders, Philadelphia.

Weiskrantz, L., 1987. *Blindsight: a Case Study and Implications.* Oxford University Press, Oxford.

Weiskrantz, L., E.K. Warrington, M.D. Sanders and J. Marshall, 1974. Visual capacity in the hemianopic field following a restricted occipital ablation. *Brain* 97, 709-728.

Weizsäcker, V. von, 1940. *Der Gestaltkreis.* Sringer, Berlin.

Weizsäcker, V. von, 1950. *Diesseits und Jenseits der Medizin.* Koehler, Stuttgart.

Weizsäcker, V. von, 1956. *Pathosofie.*Van den Hoeck und Ruprecht, Göttingen.

Weller, M., 1983. *The Scientific Basis of Psychiatry.* Baillière-Tindall, London.

Westermeyer, J., 1985. Psychiatric diagnosis across cultural boundaries. *American Journal of Psychiatry* 142, 798-805.

Wethington, E., and R.C. Kessler, 1986. Perceived support, received support, and adjustment to stressful life events. *Journal of Health and Social Behavior.* 27, 78-89.

Whitbeck, C., 1981. A theory of health. In: A.L. Caplan, H.T. Engelhardt and J.J. McCartney (eds), *Concepts of Health and Disease,* pp. 611-626.

White, L., B.Tursky and G.E. Schwartz, 1985. *Placebo. Theory, Research, and Mechanisms.* Guilford Press, New York.

Wilkes, K.V., 1980. Brain states. *British Journal for the Philosophy of Science* 31, 111-129.

Wilkes, K.V., 1984. Is consciousness important? *British Journal for the Philosophy of Science* 35, 223-243.

Wilkins, W., 1985. Placebo controls and concepts in chemotherapy and psychotherapy research. In: L. White, B. Tursky and G.E. Schwartz (eds), *Placebo. Theory, Research, and Mechanisms*, pp. 83-109. Guilford Press, New York.

Winch, P., 1987. *Trying to Make Sense*. Blackwell, Oxford.

Winson, J., 1985. *The Biology of the Unconscious*. Anchor Press / Doubleday, Gordon City, New York.

Wojciechowski, F.L., 1984. *Double Blind Research in Psychotherapy*. Swets and Zeitlinger, Lisse.

Wuketits, F., 1985. *Zustand und Bewusstsein, Leben als Biophilosophische Synthese*. Hoffman und Campe Verlag, Wien.

Wulff, H.R., 1976. *Rational Diagnosis and Treatment*. Blackwell, Oxford.

Wulff, H.R., 1986. Rational diagnosis and treatment. *Journal of Medicine and Philosophy* 11, 123-134.

Wulff, H.R., S.A. Pederson and R. Rosenberg, 1986. *Philosophy of Medicine, an Introduction*. Blackwell, Oxford.

Wüthrich, B., and T. Hofer, 1986. Nahrungsmittel-Allergien. III. Therapie: Eliminationsdiät, symptomatische medikamentöse Prophylaxe und spezifische Hyposensibilisierung. *Schweizerische Medizinische Wochenschrift* 116, 1401-1410 and 1446-1449.

Yalom, I.D., 1980. *Existential Psychotherapy*. Basic books, New York.

Young, I.M., 1984. Pregnant embodiment: subjectivity and alienation. *Journal of Medicine and Philosophy* 9, 45-62.

Zales, M. (ed.), 1985. *Stress in Health and Disease*. Brunner/Mazel, New York.

Zuriff, G.E., 1985. *Behaviourism: A Conceptual Reconstruction*. Columbia University Press, New York.

INDEX